# INSTRUMENTATION & SIGNALS

### FIELD TROUBLESHOOTING

### BOOTS ON THE GROUND
### BOOK 1

## ROBERT CUMMER

**INSTRUMENTATION & SIGNALS: FIELD TROUBLESHOOTING**

Book 1 in the *Boots on the Ground* Series

Copyright © 2026 by Robert Cummer

All rights reserved. No part of this publication may be reproduced, distributed, or transmitted in any form or by any means, including photocopying, recording, or other electronic or mechanical methods, without the prior written permission of the publisher, except in the case of brief quotations embodied in critical reviews and certain other noncommercial uses permitted by copyright law. For permission requests, write to the publisher at the address below.

**Published by:**

Mangrove Publishing LLC

801 Joe Mann Blvd

Suite P7

Midland, MI 48642

mangrovepublishing.us

**ISBN 979-8-9934979-7-6** (Paperback)

**ISBN 979-8-9934979-3-8** (eBook)

**First Edition: March 2026**

**Disclaimer:**

The information in this book is provided for educational and informational purposes only. While the author has made every effort to ensure accuracy, industrial equipment and control systems vary widely. Always follow manufacturer specifications, local codes, and your facility's safety procedures. The author and publisher assume no liability for damages resulting from the use of information contained herein.

Printed in the United States of America

# TRADEMARKS AND PRODUCT NAMES

The following trademarks and registered trademarks are the property of their respective owners and are referenced in this book for identification and educational purposes only. No endorsement by or affiliation with any trademark holder is claimed or implied.

Allen-Bradley, CompactLogix, ControlLogix, MicroLogix, PanelView, Rockwell Automation, RSLinx, RSLogix, and Studio 5000 are trademarks of Rockwell Automation, Inc.

Siemens, SIMATIC, S7, TIA Portal, and WinCC are trademarks of Siemens AG.

Modicon and Schneider Electric are trademarks of Schneider Electric SE.

Beckhoff and TwinCAT are trademarks of Beckhoff Automation GmbH & Co. KG.

Mitsubishi is a trademark of Mitsubishi Electric Corporation.

Omron is a trademark of Omron Corporation.

ABB is a trademark of ABB Group.

DeviceNet and EtherNet/IP are trademarks of ODVA, Inc.

HART is a trademark of FieldComm Group.

PROFIBUS and PROFINET are trademarks of PROFIBUS & PROFINET International (PI).

Turck, Balluff, ifm, Sick, Pepperl+Fuchs, Banner, Keyence, and Cognex are trademarks of their respective manufacturers.

Rosemount, Emerson, Honeywell, Yokogawa, Endress+Hauser, and Foxboro are trademarks of their respective companies.

All other product names, company names, and brand names mentioned in this book are trademarks or registered trademarks of their respective owners. The author and publisher make no claim to these trademarks and use them solely for reference and identification purposes.

The inclusion of any product or company name does not imply endorsement, sponsorship, or affiliation. Technical specifications, features, and product information are provided for educational purposes and may have changed since publication. Readers should consult manufacturer documentation for current specifications.

*To all the technicians and engineers with their boots on the ground— putting out fires, keeping the machines running, and showing up when things break.*

*This one's for you.*

# CONTENTS

*How to Use This Book* ... ix

Safety Fundamentals ... 1

### PART I
## WELCOME TO REALITY

1. VCR SHOP ... 11
2. Oil Fountain ... 15
3. Getting Serious ... 19
4. You're Going to Screw Up ... 21
5. Boots on the Ground ... 25
6. First Principles of Troubleshooting ... 27
7. PART I SUMMARY ... 29

### PART II
## ANTI-PANIC SYSTEM

8. The SITVD Method ... 33
9. Tools for Each SITVD Step ... 39
10. The Isolation Principle ... 47

### PART III
## KNOW YOUR MACHINE

11. Discrete Inputs ... 61
12. Discrete Outputs ... 75
13. Analog Loops ... 89
14. Temperature Sensors ... 105
15. Pressure Sensors ... 123
16. Flow Meters ... 131
17. Level Instruments ... 141
18. Load Cells & Scales ... 149
19. Photoelectric Sensors ... 157
20. Proximity Sensors ... 165
21. Cable Assemblies ... 175
22. PLC Languages & Logic ... 183

PART IV
**PUTTING IT TOGETHER**

23. Case Files ........................................................... 199
    Closing Thoughts ................................................ 219
    APPENDICES ..................................................... 223
    APPENDIX A: Pocket Math for the Field ................. 225
    APPENDIX B: Discrete Troubleshooting .................. 253
    APPENDIX C: Analog Loop (4-20 mA)
    Troubleshooting ................................................. 257
    APPENDIX D: Proximity Sensor Troubleshooting ..... 259
    APPENDIX E: Photoeye Troubleshooting ................ 261
    APPENDIX F: Cable and Wiring ............................ 263
    APPENDIX G: Temperature Sensor Troubleshooting .. 271
    APPENDIX H: The Technician's Field Kit ................ 285
    APPENDIX I: ISA Instrument Tag Quick Reference ... 301
    APPENDIX J: Common PLC Error Codes ............... 305
    Glossary of Terms .............................................. 311
    About the Author ............................................... 323

# HOW TO USE THIS BOOK

You're standing in front of a faulted machine. Production is stopped.

Your supervisor is watching. You need answers, fast.

This book is your field manual—not a textbook, not a vendor manual, but a practical guide built from thirty years of troubleshooting under pressure in automotive plants, chemical facilities, food processing lines, and assembly operations.

## WHAT THIS BOOK COVERS

This is Book 1 in the Boots on the Ground series. It focuses on instrumentation and signals: the sensors, transmitters, and field devices that tell your PLC what's happening in the real world.

You'll learn to troubleshoot:

- Discrete outputs (solenoids, contactors, indicator lights)
- Analog loops (4-20 mA signals, scaling, noise)
- Temperature sensors (thermocouples, RTDs, thermistors)
- Pressure transmitters (strain-gauge, piezo, capacitive)
- Flow meters (mag meters, turbine, Coriolis)
- Level sensors (ultrasonic, radar, float, load cells)
- Proximity and photoelectric sensors
- Cable assemblies and where they fail
- Basic PLC logic troubleshooting

## HOW THIS BOOK IS ORGANIZED

### Part I: Welcome to Reality

- Real-world pressure scenarios and the SITVD framework (Symptom, Isolate, Test, Verify, Document)—your anti-panic protocol for equipment failures.

### Part II: Know Your Machine

- Deep dives into each sensor type: how they work, how they fail, and how to fix them fast.

### Part III: Putting It Together

- Case files from real troubleshooting situations. These aren't sanitized textbook examples—they're messy, pressure-filled problems that teach you to recognize patterns.

### Appendices

- Pocket math, decision trees, quick-reference charts. Photocopy these, laminate them, keep them in your toolbox.

———

## HOW TO USE IT IN THE FIELD

When you're troubleshooting:

1. Start with the decision trees (Appendix E)—they'll walk you through the SITVD isolation chain
2. Reference the relevant chapter for detailed procedures
3. Use the pocket math (Appendix A) to verify readings make sense
4. Check the case files (Chapter 14) to see similar problems solved

When you're learning:

- Read it front to back. The pressure scenarios in Part I hook you, the technical chapters in Part II teach you, and the case files in Part III show you how it all comes together.

## This book is NOT:

- A vendor product manual (check those for specs)
- A comprehensive theory textbook (we skip the math you don't need)
- A wiring diagram reference (that's machine-specific)
- A substitute for proper training (use it to supplement, not replace)

## This IS:

- A field guide to systematic troubleshooting
- A pattern recognition tool built from experience
- A confidence builder when the pressure is on

## TERMINOLOGY & CONVENTIONS

⚡ Shop-Floor Wisdom = Hard-won lessons from the field

💡 Pro Tip = Practical advice that saves time

⚠️ Critical = Safety or major issue—pay attention

Volts, amps, ohms use standard abbreviations (V, A, Ω)

Pressure units: PSI (pounds per square inch)

Temperature: °F unless noted

Flow: GPM (gallons per minute)

## FREE RESOURCES

Full-size printable decision trees, troubleshooting checklists, and bonus material available at:

www.BootsOnTheGroundTech.com

Enter code: COLDSPRAY to get access.

## NOW LET'S GET TO WORK

Every complex system is just a chain of smaller, solvable problems.

Your job is to break the chain down, find the broken link, and fix it.

This book will teach you how.

Turn the page and let's start troubleshooting.

# SAFETY FUNDAMENTALS

## SAFETY FUNDAMENTALS—READ THIS FIRST

I've seen a lot of industrial accidents over thirty years. Most of them could have been prevented.

We all strive for zero OSHA recordables, but things do happen. Equipment fails, procedures get skipped under pressure, and people make mistakes. I opened this book with a close call that could have resulted in slip-and-fall injuries, trips to the eyewash station, or worse. We got lucky that time. But luck isn't a safety strategy.

Whether you're working in a factory or a chemical plant, you must follow all safety guidelines and rules. This book teaches troubleshooting, but troubleshooting safely is just as important as troubleshooting correctly.

If you are not qualified to perform any of the duties outlined in this book, seek the appropriate qualified or electrically certified individual. There's no shame in saying, "I need help with this." There's plenty of shame in getting hurt because you didn't want to ask.

This guide provides field-practical diagnostics for isolating common cable failures, from conductor fatigue to insulation degradation. It is organized sequentially to prioritize technician safety while systematically identifying the most probable points of failure, enabling efficient repair or replacement in time-critical environments. **However, efficiency never trumps physical safety.**

## LOCKOUT/TAGOUT (LOTO) IS NON-NEGOTIABLE

If you're working on equipment that could move, energize, or release stored energy, you lock it out. Every time. No exceptions.

Lockout/Tagout means:

- De-energize the equipment (turn off power, close valves, release pressure)
- Lock the energy isolation points with your personal lock
- Tag the equipment so everyone knows it's being worked on
- Test that it's actually de-energized before you touch anything

I don't care if "it'll only take a second" or "production is breathing down your neck." Lock it out. Production can wait five minutes. Your fingers can't grow back.

**When you DON'T lock out:**

- Diagnostics that require the machine to be energized (metering voltages, watching PLC logic online, checking sensor outputs)
- For these tasks: Use proper PPE, stay clear of moving parts, and never reach into energized equipment

In this industry, your reputation is built on your technical skill and your safety record, not on how many corners you were willing to cut. If the equipment isn't locked out, the work doesn't start. It is a fundamental boundary of the trade that everyone—from the floor to the front office—must respect.

There is no "quick fix" or "minor adjustment" significant enough to justify the risk of an unexpected start-up. When you refuse to work unsafe, you aren't being difficult; you are being a professional. You are protecting yourself, your coworkers, and the equipment.

# ELECTRICAL QUALIFICATION MATTERS

You'll be working with voltages ranging from 24 VDC control circuits to 480 VAC motor power. Not all of these require the same level of training or qualification.

**Know your limits:**

- If you're not electrically qualified for the voltage level you're working with, stop
- Call someone who is qualified
- Arc flash hazards at 480V can kill you from several feet away—you don't have to touch it to get hurt

**What's typically acceptable for instrument techs** (with proper training):

- Metering 24 VDC control circuits
- Checking sensor outputs at PLC terminals
- Viewing PLC logic online
- Replacing sensors and field devices (power locked out)

**What requires qualified electricians:**

- Working inside energized electrical panels
- Working on 120 VAC or higher circuits
- Installing or replacing motor starters, VFDs, or power distribution equipment
- Any work requiring arc flash PPE beyond basic safety glasses and gloves

Know your facility's rules; what's allowed in one plant may be forbidden in another. When in doubt, ask. It is better to admit you aren't qualified than to explain an accident.

## STORED ENERGY KILLS

Even when power is off, energy can still be stored in the system:

- Compressed air in cylinders or lines
- Hydraulic pressure in accumulators
- Springs under tension
- Capacitors in power supplies or VFDs
- Rotating mass (flywheels, fans, conveyors)

Before working on anything, ask yourself: "Where is energy stored in this system, and how do I safely release it?"

Bleed air pressure. Discharge capacitors. Release spring tension. Block moving parts. Then—and only then—start working.

## NEVER BYPASS SAFETY INTERLOCKS

Safety interlocks exist because someone got hurt in the past. Guard door switches, light curtains, two-hand controls, e-stops—they're not suggestions.

When troubleshooting, you might be tempted to jumper out a guard switch to see if that's the problem. Don't.

**Safe ways to test interlocks:**

- Go online with the PLC and check the input bit
- With the machine locked out, jumper at the PLC terminal to verify wiring
- Follow your facility's lockout defeat procedures (if they exist)

If you can't figure out why an interlock is preventing operation, that's when you escalate—not when you bypass it and hope for the best.

# HAZARDOUS LOCATIONS CHANGE EVERYTHING

This book does not explicitly cover working in hazardous (classified) areas —Class I, Division 1 or 2 environments where flammable or explosive vapors, gases, or dusts are present.

I have many years of experience working directly with and around hazardous materials. Working in these environments changes the game completely. You must know:

- What materials you're working near (check MSDS sheets)
- What classification zone you're in (Class I Div 1 vs. Div 2 vs. unclassified)
- What intrinsically safe (IS) barriers are in place and how they work
- What hot work permits are required
- What monitoring equipment is required (LEL meters, gas detectors)

**If you are at all unsure of yourself or haven't reviewed the Material Safety Data Sheets for the area you're working in—don't be a hero. Stop and get help.**

Flammable vapors don't give you a second chance. One spark in the wrong place can kill not just you, but everyone around you.

If your plant has classified areas, you've already been through specialized training. Follow it. If you haven't been trained for hazardous locations, don't work in them until you are.

# KNOW WHEN TO ESCALATE

Some situations are beyond field troubleshooting. They require specialized training, equipment, or procedures:

- High-voltage work (>600V)
- Confined space entry
- Hazardous atmosphere work
- Hot work permits (welding, cutting, grinding)
- Working with cryogenic fluids or extreme temperatures
- Work requiring arc flash PPE you don't have

If you're not trained and equipped for the situation, call someone who is. There's no shame in escalating. There's plenty of shame in creating an incident because you thought you could handle it.

## TEST EQUIPMENT SAFELY

Your meter is essential for troubleshooting, but use it correctly:

- Inspect meter leads for damage before every use
- Set the meter to the correct function (V, A, Ω) and range
- Never meter resistance on live circuits—you'll blow the fuse or damage the meter
- Use proper category-rated test leads (CAT III minimum for industrial)
- When metering high voltage, keep one hand in your pocket (reduces current path through your heart)

## PRESSURE IS NOT AN EXCUSE

This book emphasizes troubleshooting under pressure. The line is down, production is stopped, supervisors are asking for time estimates, and everyone's watching.

I've been there hundreds of times. The pressure is real.

But here's the truth: **That pressure is never an excuse to skip safety procedures.**

No supervisor will thank you in the hospital for skipping lockout to save five minutes. No production manager will cover your medical bills because you bypassed an interlock. And no machine is worth dying over.

When you're standing in front of a faulted machine and the pressure is on:

1. Take a breath
2. Follow lockout procedures if required
3. Verify de-energized before touching anything
4. If you're not qualified for the task, get someone who is
5. Think before you act

The SITVD framework in this book is designed to help you think systematically under pressure. Use it. Don't let pressure override your training.

## DOCUMENT SAFETY ISSUES

If you find a safety problem during troubleshooting—missing guards, bypassed interlocks, damaged wiring, leaking hydraulics—document it and report it immediately.

- Tag it out if it's an immediate hazard
- Notify your supervisor
- Fill out a safety report or work order
- Don't assume "someone else will take care of it"

You found it. You own it. Fix it or make sure it gets fixed.

## FINAL WORD

I wrote this book to help you troubleshoot faster and with more confidence. But no amount of speed is worth getting hurt.

Your greatest skill as a troubleshooter is solving problems under pressure. Your greatest responsibility is going home safe at the end of your shift.

# PART I
# WELCOME TO REALITY

# CHAPTER 1
# VCR SHOP

The VCR Shop—Where it all started

My introduction to troubleshooting didn't happen in a chemical plant or with a million-dollar PLC. It happened in a little VCR repair shop in the late 1980s. Back then, movies came on VHS tapes, and people brought their machines in when they broke down instead of throwing them away.

I was still in high school, dual-enrolled in a vocational program, and lucky enough to get a co-op job at a local VCR repair center. My job was simple: take the complaint, pull the cover, and start checking. Back then, manufacturers shipped schematics with their products. You could actually follow the circuit diagrams and figure out where the fault was. Today, most consumer electronics are sealed shut and discarded when they fail. That wasn't the case then.

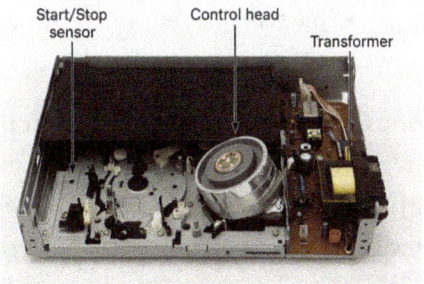

*Image Interior view of common VCR chassis*

One of the most common problems in a VCR was a bad start or stop sensor. These were usually simple infrared LED/phototransistor pairs—optical interrupters mounted where the tape threaded into the carriage. Their job was to confirm tape movement: the machine needed to know when the tape had started and when it had reached the end of the reel.

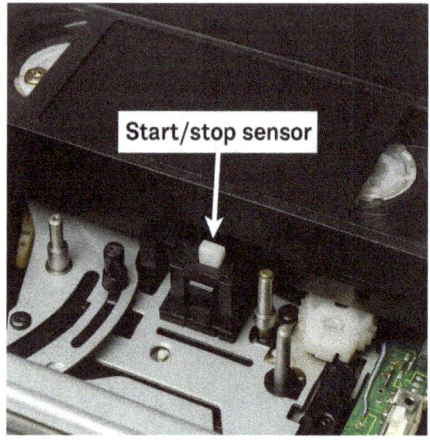

*Image Close up view of common VCR Start/Stop Sensor*

When a start/stop sensor failed, the logic got confused. The tape would pull in, then eject immediately, or the machine would refuse to play at all. The sensor might test fine when cold, but then drift out of specification once it heated up. That's where the cold spray came in—blast it, cool it, and if the signal came back, you had your culprit. Today, Freon is no longer used, but the principle remains the same: temperature stress is a quick way to expose weak or failing components.

There were also "hands-on" checks. Blocking the light path with your finger should have registered an "off" state. If the system didn't respond, the sensor or its wiring was suspect. It was crude but effective—and honestly, some of the best troubleshooting still comes down to quick, common-sense tests like that.

Another common problem was the control head. That's the spinning drum with the magnetic read heads, the equivalent of a phonograph needle. A dirty head would scramble the picture or altogether disable playback. You had to clean it carefully with isopropyl alcohol or an approved electronic cleaning solution—nothing that left residue. And you couldn't just use a cotton swab; those could snag or leave fibers. You used lint-free pads, careful strokes, and patience. Even here, common sense and discipline mattered.

Looking back, I realize these small jobs were my first exposure to a universal troubleshooting pattern that I've carried into every job since:

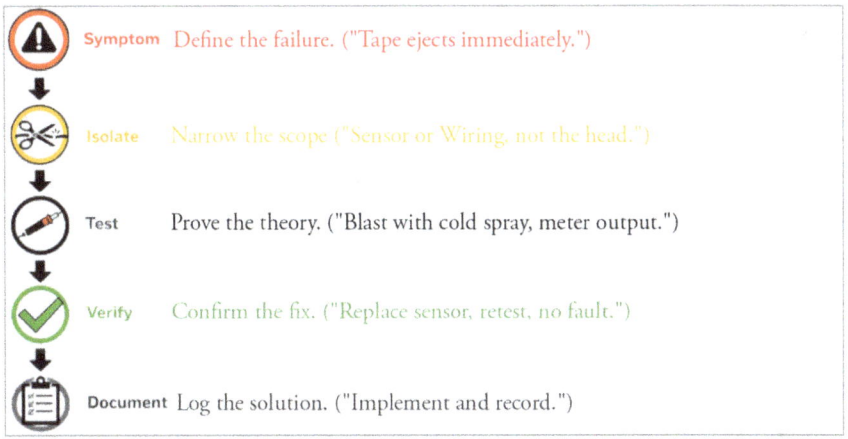

## SITVD: THE PATTERN THAT SCALES

In 30 years of troubleshooting, from VCR repair shops to auto plants and chemical plants, I've learned one thing: every problem follows this same pattern. It's the same workflow I use today on a 4–20 mA loop in a refinery. It doesn't matter whether you're fixing a $200 consumer device or a $200,000 flowmeter. The mindset and the sequence are identical.

Master this pattern, and you'll fix nearly anything:

Symptom, Isolate, Test, Verify, Document

## Cockroaches and Cold Spray

And then there was the day a customer brought in a VCR wrapped in a plastic grocery bag. The complaint: "It's jammed."

I popped the cover, and all hell broke loose. Dozens of cockroaches poured out of the machine. They had multiplied inside the warm, dark chassis. The bench, the floor, even my lap—crawling with bugs. We had to fumigate and bug-bomb the shop to keep working.

That was my first lesson in troubleshooting: not every problem is clean, not every fix is glamorous. Sometimes you start by clearing the cockroaches. But once the mess is out of the way, the pattern is the same: symptom, isolate, test, verify, document.

The cockroach VCR taught me something else too: troubleshooting isn't just about electronics or mechanics or software. It's about staying calm when chaos erupts, thinking clearly when everyone else panics, and following a systematic approach even when you're covered in bugs.

This book is about that pattern—the one that works whether you're knee-deep in cockroaches or standing in front of a million-dollar production line that just went down. It's about the real world of troubleshooting, where problems are messy, time is money, and the only thing that saves you is a methodical approach and the discipline to follow it.

Every story in this book, every technique, every tip—they all come back to the same five steps. Some days you'll use fancy tools and sophisticated diagnostics. Other days, you'll use cold spray and your finger. But the pattern never changes.

Let me show you how it works.

One more thing before we move on, enter 'COLDSPRAY' at https://bootsonthegroundtech.com/ to access downloads and bonus features.

# CHAPTER 2
# OIL FOUNTAIN

**The Oil Fountain**

Fast-forward a few years. My first real job out of college was at an off-site test lab for AC Rochester, later called ACDelco. I was an automotive test engineer, working on the validation of lube oil and fuel filters. It was dirty, smelly work—the opposite of what I thought I'd be doing with my electrical engineering degree. However, it was real engineering, and it taught me lessons that have stuck with me for life.

At the time, gaining entry to General Motors in Flint was no small feat. This lab—AC Rochester, later ACDelco—was essentially GM's research and validation arm. We weren't just building rigs for fun; we were testing lubrication oil and fuel filters that directly tied into warranty claims, lawsuits, and customer complaints.

Car dealers would send us filters from failed engines and claim they had blown out or caused knocks. Half the time it was real, half the time it was bullshit—either the filter wasn't changed, or the dealership wanted to pad a warranty claim. We'd cut the filters apart, test anti-drain back valves, measure filter media capacity with dirt load, and even run gravimetric particle analysis. If we wrote the wrong conclusion, it could end up in court.

It wasn't glamorous. I had wanted to design audio amps, not scrape oil and smell like fuel. I was making under nine bucks an hour after college.

However, this was research and development, and it marked the beginning of my career. Looking back, it was a hell of a training ground. The oil fountain disaster came right out of that world.

One of my first assignments was to design and build a pressure impulse test stand. Picture a tank of oil with a pump bolted to the top, hoses snaking into a directional control block, and a small PLC cycling it all back and forth. It wasn't elegant, but it worked. The lab itself smelled like hot oil and solvents, a strange mix of old and new equipment humming along, and maple-top benches covered with tools and stained paperwork. I was fresh out of school, proud of the Frankenstein rig I'd stitched together, and more than a little eager to prove myself.

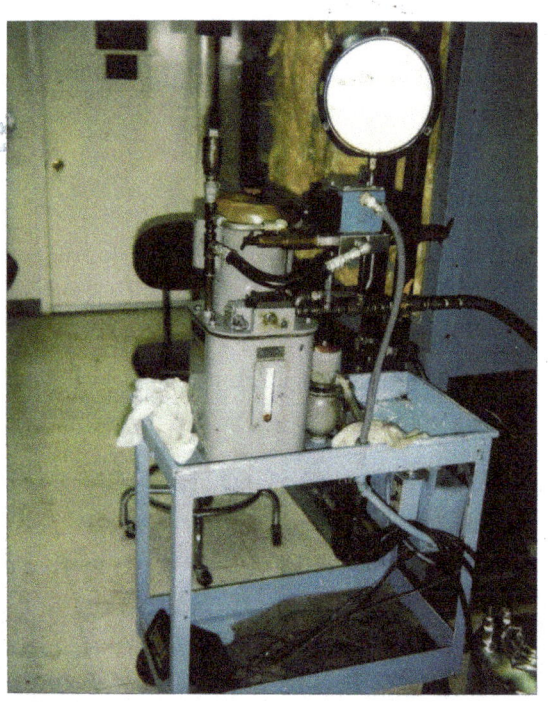

*Image Pressure Impulse Stand built by the author in Flint, Michigan, May 1995.*

We wired it, tightened fittings, purged air out of the lines. Everyone stood around, quiet, waiting for me to hit the button. My coworker flashed me a thumbs-up.

**Big mistake!**

I immediately pressed Start, and in an instant the place turned into the oil-lab version of Yellowstone National Park. A 40-gallon-per-minute geyser of 10-weight oil shot straight out of an uncapped port and hammered the ceiling sixteen feet overhead. It coated fluorescent lights and steel beams, then rained back down like a storm. Within seconds, everyone was drenched. The stand, the desks, the paperwork, the floor—everything had a glossy new finish courtesy of me.

We were laughing, but it was nervous laughter. I'd basically built Busch Gardens in a lab bay. We dragged out mops, buckets, Pig-blankets and rags, but it didn't matter. The oil had soaked into everything. For years afterward, the stains remained. Every so often, a drop of oil would still fall from above as a reminder of that day.

Here's the thing: it was *funny*—but only because nobody got hurt. We were an off-site lab, so management didn't come down on us, but we weren't really being safe. A jet of oil under pressure can do worse than make a mess; it can burn skin, blow out eyes, or create slip hazards that wreck somebody's back. Being young wasn't an excuse, because experienced guys were standing right there with me. And even they didn't catch it.

That was my first *real* lesson: never trust a thumbs-up. Walk the skid yourself. Trace every port, every valve, every fitting. Don't assume "all good" means you're safe. And above all, remember that experienced people make mistakes too. If you don't check for yourself, you'll pay for it.

Humility came quick. I walked into that lab with a degree and a sense that I was ready for anything. Five minutes later, I was covered in oil, looking like a fool in front of everybody. But that sting turned into discipline. I started double-checking, triple-checking, and refusing to push Start until I understood the system completely.

Every plant engineer / tech has a story like this. It might be a miswired sensor, a breaker left off, or in my case, an oil geyser that turned the lab into a water park. You will screw up. It'll be public. You'll feel like an idiot. But if you take it seriously, you'll never make the same mistake twice.

The Oil Fountain was mine. It humbled me, and it made me sharper. I learned the hard way that troubleshooting isn't just about fixing—it's about preventing, about being disciplined enough to check before you commit. That mix of humor and pain was my initiation into the reality of engineering: this job is unforgiving, but if you respect it, it'll make you better every single day.

### ⚠️ *Shop-Floor Tip*

*Pressure is unforgiving. Even a "small" port left open can create a jet strong enough to injure or blind. Before you ever press Start, walk the system yourself. Trace every port, every valve, every fitting. Never rely on someone else's thumbs-up.*

# CHAPTER 3
# GETTING SERIOUS

## GETTING SERIOUS

### (or: Why Anyone Should Waste Time Teaching You)

The oil fountain story is funny now. It wasn't funny then. But here's what I need to tell you that nobody wants to admit in a technical book: *I wasn't always the guy you'd want to invest time in teaching.*

Early in my career at that test lab in Flint, I had my two-year degree and a job title. I thought that was enough. I'd show up, do what was asked, and go home. When equipment broke, I'd take the easy route—call someone else, wait for help, avoid the hard thinking. I wasn't invested in actually *learning* how things worked. I was just putting in time.

Then something clicked. I realized that the experienced techs and engineers around me weren't going to waste their time teaching someone who wasn't serious about learning. Why would they? They had their own work to do. If I wanted to actually *know* this stuff—to be the person people called when things broke, not the person waiting for that call—I had to prove I was worth the investment.

So I started showing up differently. Staying late to understand why a machine failed instead of just getting it running again. Asking real questions instead of nodding along. Actually reading the manuals instead of guessing. Tracing circuits on prints until I understood the *why*, not just the *what*.

The skills were always there. I just hadn't unlocked them yet because I hadn't taken it seriously.

**That's why this book exists—and why I'm telling you this.**

If you bought this book, you're already doing the work. You're looking for the system. You're willing to invest in yourself. That means you're serious, even if you don't feel confident yet.

Now let me show you what I learned the hard way: **the framework that keeps you methodical when everyone else is panicking, and the discipline that separates techs who survive from techs who thrive.**

In the coming chapters, we are going to strip away the guesswork and look at the fundamental physics of how these systems actually fail. We'll cover the mindset and the tools first, because a framework is useless if you don't understand the ground you're standing on.

Once that foundation is set, I'm going to hand you the **SITVD method**. It is the same systematic workflow I used to survive those early years, and once you internalize it, you'll use it every single day for the rest of your career.

# CHAPTER 4
# YOU'RE GOING TO SCREW UP
## (AND WHY THAT'S GOOD)

I'll just say it plainly: you will screw up. Everyone does. It doesn't matter whether you're a fresh co-op straight out of school or a seasoned contractor with decades under your belt. If you work in this trade long enough, you're going to have your "what the hell just happened?" moment. The trick isn't avoiding mistakes altogether—that's impossible. The trick is learning which kind you're making, what they cost you, and how to make sure you never repeat the critical ones.

## THE THREE CATEGORIES OF MISTAKE

Over the years, I've learned there are really three categories of mistakes you'll run into. Each carries its own weight, but they all serve as the "scars" that eventually make you an expert.

### 1. Embarrassing but Harmless

These are the mistakes that bruise your ego more than anything else. Maybe you swap two wires and wonder why nothing works. Maybe you loaded the wrong program revision and spent an hour chasing ghosts before you realized it. Or maybe you hit Start and end up with an oil geyser that turns a whole lab into Busch Gardens. Nobody got hurt. Equipment isn't ruined. But you feel like an idiot. And here's the good part: those moments stick. You'll laugh about them later, but at the time they sear the lesson into your brain in a way no training manual ever could. They make you humble—and humility on the job is a gift.

## 2. Critical Safety Mistakes

These are the ones you cannot afford. Miss a lockout, trust a thumbs-up instead of testing voltage, assume a line is depressurized when it isn't—these aren't funny stories later. They're the mistakes that can put someone in the hospital or worse. Nearly everyone thinks they won't make safety mistakes, but pressure, fatigue, and bad habits creep in. You don't rise to the occasion; you fall back to your training. If your default habit is to verify isolation, test before touching, and respect pressure, you'll be fine. If it isn't, you'll eventually pay the price. The good news is you can learn from embarrassing mistakes early, so you don't graduate to safety mistakes later.

## 3. Critical System & Business Mistakes

This third category doesn't get talked about as much, but it can be just as brutal in its own way. These are mistakes that don't physically hurt anyone but can cost a plant hundreds of thousands in downtime or make you look like a fool in front of management and the union hall. I learned this one the hard way while contracting through my company, Iris Automation. I was brought into an auto plant to troubleshoot an end-of-line test cell. Like most modern lines, the process was broken into "cells"—assembly, welding, adhesive application, final assembly, then test. The assumption—mine included—was that the cells were modular. If I shut one down, the others could keep working.

That assumption cost the plant time and money!

Nobody told me this particular cell wasn't modular. It was serial. Worse, it was also the local SQL database server feeding test data into the ERP system. When I shut it down, I didn't just pause my machine—I cut off the data backbone for the entire line. Suddenly the rest of production didn't know what assemblies were coming next. The line stopped cold. Within minutes, the plant supervisor was storming down, radios were crackling, and the union foreman was chasing down piece counts because workers couldn't be released without them. Thirty machines sat idle because I'd shut down one. Technically, it wasn't "my fault." I had an escort, and even the in-house engineer didn't know about the dependency. But here's the lesson: production doesn't care whose fault it is. The plant was down, and I was the guy who had pushed the button. What should have happened was a huddle before touching the machine: stakeholders aligned, risks called out, plan agreed. Instead, I became the face of a critical business mistake.

## What to Do With Your Mistakes

So yes—you will screw up. Sometimes it'll be private, sometimes public. Sometimes harmless, sometimes serious. The important part happens next.

---

**Own it**. Don't hide. Don't point fingers. You'll earn more respect by admitting the screw-up than trying to dodge it.

**Learn from it**. Every mistake is a deposit into your mental troubleshooting playbook. Pay attention to the lesson or you'll pay for it again.

**Don't repeat it**. One oil geyser is funny. Two make you careless.

**Respect safety**. If you're going to be wrong, be wrong in a way that nobody gets hurt.

---

And finally, when it's somebody else's turn, don't pile on. Everyone has their war stories. The fact that you've survived yours and can pass the lesson forward is what makes you valuable on the floor or in the field.

## The Long View

Here's the good news: You will screw up less over time. The longer you work, the better you get at spotting traps before they spring. You'll hesitate before pressing Start. You'll double-check that breaker. You'll think twice about shutting down a machine without knowing what it's tied into. The mistakes never vanish entirely—but they get smaller, less frequent, and less costly.

That's the whole point of this section, and really of this book: mistakes are not the end. They're the feedback loop that makes you sharper, safer, and more disciplined. If you're willing to learn from them, they'll be the best training you'll ever get.

### ⚠ Three Categories of Mistakes

1. **Embarrassing but Harmless** You'll feel stupid, but nobody gets hurt. These sting your pride and make great war stories.
2. **Critical Safety Mistakes** The ones that can injure or kill. These must be eliminated through discipline and habit.
3. **Critical System/Business Mistakes** Shut down a line, stall production, or kill data flow. Nobody gets hurt, but your credibility does.

---

👉 *Learn from the harmless ones early so you never graduate to the dangerous ones.*

---

# CHAPTER 5
# BOOTS ON THE GROUND

Every plant environment is different. Some are dirty and gritty, with dust in the air and oil on the floor. Others are spotless, organized, and quiet enough to hear your own footsteps. The point is: be prepared for anything. If you're on your own, you learn quick. If you're a contractor walking into a new plant, you really learn quick.

On any given day, you'll deal with fork trucks buzzing past, beepers and horns going off, overhead PAs calling out shift changes. You'll hear clicks and clacks that sound like your machine failing—but it's just the hum of the plant. And if the place is quiet? That usually means something's wrong, because production doesn't stop unless there's trouble.

Sometimes the challenges are simple but frustrating. Panels hidden behind equipment that was built around them decades ago. Crawling on the floor through old gum and grime just to get a line of sight on a conduit. Standing at the parts crib waiting for someone on lunch while the machine you're supposed to fix sits down hard. Nothing in a classroom prepares you for those moments. You only learn by living them.

And then there are the hazards. Poor air exchange in paint booths. Confined spaces. Vibration and noise that numb your senses. Sometimes the "work environment" itself becomes the biggest part of troubleshooting. You'll sweat, and not just from the heat—but from nerves, because when the line is down and you're the rookie, it feels like the world is watching. You panic: I'm going to get fired. I'm supposed to have this fixed already.

That's when the basics matter most. Keep calm. Rely on your fundamentals. Communicate with the team. Nine times out of ten, that's what carries you through.

I learned this the first time I chased what looked like a PLC output fault. The program looked fine, the output light was on, but the valve never fired. My first instinct was to blame the card. I was ready to pull hardware when I finally noticed the junction box hanging off the side of the machine. Vibration had loosened a screw terminal just enough to break contact. Tightened it down, problem solved. A half-hour of stress, all because a $0.02 screw worked loose.

That was my first mental decision tree.

---

Step one: don't assume.

Step two: check the simple stuff before you escalate.

Step three: use your eyes and hands, not just the screen.

---

And here's the part you'll learn quickly: as long as you've got two things—your set of prints and a multimeter—you're going to be okay. That's your machine bible. With those two tools, you can trace, verify, and isolate almost any problem. Everything else builds from there.

So don't kid yourself. Troubleshooting isn't just typing on a laptop. It's dirt, noise, sweat, and sometimes panic. But it's also the satisfaction of finding the loose wire, tightening the terminal, and watching the line come back to life. Out here, the basics win—*boots on the ground*.

# CHAPTER 6
# FIRST PRINCIPLES OF TROUBLESHOOTING

## THE TROUBLESHOOTER'S MANIFESTO

These principles carry from a cockroach-filled VCR to a multi-million-dollar plant DCS. Every problem you encounter will eventually come back to these six basics.

## THE SIX CORE PRINCIPLES

### 1. Understand the Function

- Before touching any hardware, ask: *What is this device supposed to do?*
- A valve is meant to open and close; a sensor is meant to measure exactly one variable. If you do not know the intended function, you cannot identify when it is failing.

### 2. Know the Failure Modes

- Learn the "usual suspects": proximity sensors losing alignment, RTDs picking up noise, or contractors burning out.
- Mastering common failure patterns is the only way to increase troubleshooting speed.

## 3. Test and Verify

- Avoid exotic theories. Start with the simplest tools: your multimeter, a can of cold spray, or a physical check for heat.
- The basics catch the vast majority of faults before you descend into a "rabbit hole".

## 4. Never Assume

- Assumptions are the primary cause of professional failure.
- If an operator says a valve is open, check it yourself. If a print says a wire is on terminal 17, trace it. If a PLC light is on, meter the terminals.

## 5. Stay Calm Under Pressure

- When alarms are blaring and management is pacing, panic leads to bad calls.
- Slow down, breathe, and work your decision tree; the machine will wait for you to think.

## 6. Learn From Every Mistake

- Whether the mistake is yours or someone else's, every failure is a lesson.
- If you do not change your troubleshooting workflow after a failure, you have wasted the lesson.

---

👉 Tape these to your toolbox. Burn them into your head. When downtime is ticking by the minute, these rules keep you moving.

---

# CHAPTER 7
# PART I SUMMARY

**Part I Summary: The Mental Framework**

In Part I, we focused on cultivating the right mindset for mastering troubleshooting control systems. This isn't just about the tools in your bag; it is about the orientation of the person holding them. Success in this field is built on three pillars: **humility, discipline, and a commitment to the fundamentals.**

**The Mental Framework at a Glance:**

> **Humility:** To avoid the "Assumption Trap" and remain observant.
>
> **Discipline:** To maintain a methodical workflow under intense production pressure.
>
> **Fundamentals:** To rely on physics and logic rather than guesswork and luck.

**Humility** is your primary safeguard against the "Assumption Trap." It reminds us that no matter how many years we have spent in the field, there is always a detail we might miss. Overconfidence is the quickest way to become blind to the obvious, leading to hours of wasted time chasing ghosts when the real problem was sitting right in front of us. To stay humble is to stay observant.

**Discipline** is what drives a consistent, methodical workflow. When the plant is down and the pressure is mounting, the temptation is to move with haste and take shortcuts. Discipline is the internal voice that forces you to execute every step—from initial setup to final verification—with deliberate care. It is the difference between a "quick fix" that fails an hour later and a permanent solution that holds.

Finally, a **commitment to the fundamentals** keeps your foundation solid. It means mastering the core principles of control theory, understanding the physics of system behavior, and internalizing safety protocols before you ever dive into the complexity of a 4,000-line program. Without these basics, you aren't troubleshooting; you're just guessing.

Together, these qualities form the mental framework that underpins everything you will do moving forward. They prepare you to approach any challenge, no matter how daunting, with a sense of clarity and professional confidence. You have the mindset; you have the discipline. Now, it's time to move from the philosophy to the field.

**Let's look at the signals.**

# PART II
# ANTI-PANIC SYSTEM

# CHAPTER 8
# THE SITVD METHOD

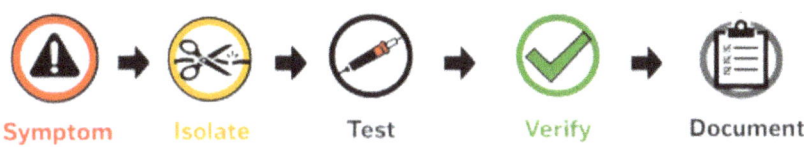

## SYMPTOM

What's Actually Wrong?

The symptom is what the machine is doing (or not doing). Not what the operator thinks is wrong. Not what happened last time. What you can observe right now.

The key is being specific. A good symptom points you toward the problem. A bad one sends you in circles.

Real example: An operator calls: "The glue station is down again." That's not a symptom. You get there and observe: "Photoeye 2 is triggering with no box present." Now you have something to work with.

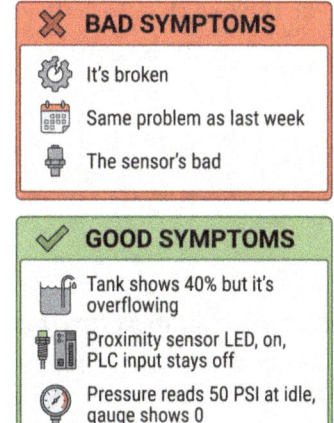

*Image Bad vs Good Symptoms*

## ISOLATE: WHERE IN THE CHAIN?

Every sensor signal follows a chain:

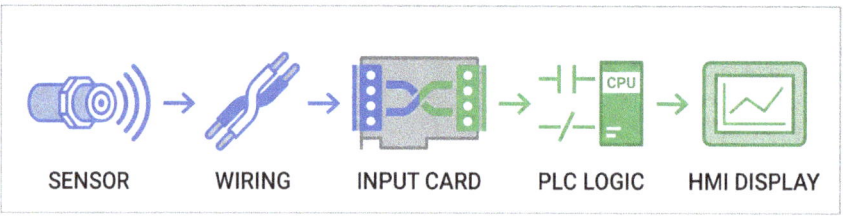

Your job is to find which link is broken. This is where most time is wasted or saved. Skip this step (like I did with the oil fountain) and you'll chase ghosts for hours.

Trace the signal backwards: from the display you can see to the element you can't.

## THE ISOLATION RULES

**Start where it's easy to test.** Don't climb the ladder to the sensor if you can check the terminal block at eye level first.

**Cut the problem in half with each test.** Identify the midpoint of the signal chain. If the signal is good there, you've just eliminated 50% of the possible failure points in one move.

**Never assume—prove each link.** Until your meter shows voltage or your software shows a bit change, that link is still a suspect.

**Example: Temperature reading is wrong.**

| Step | Discovery | Action/Result |
|---|---|---|
| HMI | Shows 200°F | Value is "live" but incorrect. |
| Logic | 16,383 Raw Counts | PLC is receiving a full-scale signal. |
| Input Card | Signal is stable | Responding to changes; card is healthy. |
| Sensor | 138.5Ω Resistance | Expected 100Ω for 32°F. |

*Table Isolation Rules*

---

**Isolated**: Bad RTD

---

## TEST: PROVE YOUR THEORY

Once you've isolated the likely problem, test it. Don't guess. Don't assume. Test.

Common test methods:

- Substitution: Swap with known good part
- Simulation: Inject a test signal (12mA = 50% for 4-20mA)
- Bypass: Jumper around suspected component
- Stimulation: Ice bath for RTD, magnet for proximity sensor

Example: You've isolated a suspicious pressure transmitter.

- Disconnect transmitter
- Inject 12mA with loop calibrator
- PLC should show 50%
- If it does → transmitter is bad

- If it doesn't → wiring or card problem

## VERIFY: CONFIRM THE FIX

The worst callback is for the same problem. That's why you verify.

Verification means:

- Run multiple full range test cycles
- Check for side effects
- Watch **for** intermittent values

Example: You replaced a proximity sensor.

- Run 10 machine cycle tests with different part types
- **Verify** counting is accurate
- Check that downstream sequences work

## DOCUMENT: CLOSE THE LOOP

Documentation isn't paperwork—it's leaving breadcrumbs for the next tech (maybe you).

### What TO document:

- What failed and why
- Part number of replacement
- Any configuration (NPN/PNP, 0-100 PSI range)
- Weird symptoms that might return

### What NOT to document:

- "Replaced sensor" (useless)

- Novel-length theory (nobody reads it)
- Blame ("operator error")

**Good documentation:** "Prox sensor PE-123 failed—internal LED stuck on. Replaced with same P/N. Set to Light-On. 2mm gap verified. Caused by coolant ingress through damaged cordset."

That tells the next tech everything they need.

## THE CHAIN IN ACTION

Here's how SITVD handles a real problem:

06:00 AM: Conveyor won't start.

| Phase | Action Taken | Action |
|---|---|---|
| Symptom | Identify the failure. | Start button is unresponsive; conveyor motor fails to energize. |
| Isolate | Define the boundary. | HMI shows "System Not Ready." PLC Rung 45 shows PhotoEye_14 (PE14) is not made, despite LED being ON. |
| Test | Prove the theory. | Blocked PE14 with cardboard—LED toggles, but PLC bit stays OFF. Jumped terminals at card—PLC sees signal. Fault isolated to field wiring. |
| Verify | Confirm the fix. | Located and repaired broken wire at flex point. Added service loop. Ran 20 cycles; re-tested all safety functions. |
| Document | Record the truth. | Logged: "PE14 wire broken at robot flex point. Re-terminated with service loop. Plan cable replacement." |

*Total time: 45 minutes. Without SITVD, this is a 3-hour wild goose chase.*

## WHY THIS WORKS

👉 Machines are stupid. They do exactly what their inputs tell them. When something goes wrong, there is always evidence—you just need to look in the right place, in the right order.

👉 SITVD isn't magic; it's discipline. It forces you to remain methodical when everyone else is panicking, keeping you from jumping to conclusions when the pressure mounts. Master this pattern on simple sensors, and you'll use the same approach on million-dollar systems. The scale changes. The pattern doesn't.

# CHAPTER 9
# TOOLS FOR EACH SITVD STEP

## PEOPLE ARE TOOLS TOO

Before we get into multimeters, loop calibrators, and scopes, let's talk about the tool most folks forget: people.

Operators, maintenance techs, electricians, and engineers in a plant are just as important to your troubleshooting kit as any piece of gear. In fact, they're often more valuable than your multimeter, oscilloscope, or laptop. Ignore that, and you'll spend twice as much time working alone.

The first thing I do when I walk into a facility is introduce myself. I make it clear: this is their house. I'm here to help, not take over. That little bit of respect goes a long way. People want to know who you are, what you're there for, and whether you'll treat them as part of the solution.

But here's the reality: the plant is *Baskin-Robbins*—you'll meet every flavor of person. Some are welcoming, some are threatened, some are too busy to care. And everyone remembers the "last guy." Maybe he cut corners. Maybe he pissed people off. Maybe he never got invited back. Listen closely when those stories come up—they're inside intelligence, and they help you avoid stepping on landmines before you even touch a panel.

You also have to read the room. Asking, "What do you think it might be?" can build trust—or it can get you the sharp reply, "That's why we hired you." That's shop talk, and it comes with the territory. The trick is to ask the right questions: When did it start happening? How often does it happen? Did anything change before it started? Those questions get people talking about facts, not speculation.

And here's the nuance they don't teach you: you can be the most pleasant, kind person in the world and still have a rough time in a plant. This environment demands a little grit. Small talk helps—maybe it's hunting, maybe sports—but there's a line between being personable and getting distracted. The human element matters, but so does getting the job done.

Sometimes the balance tips the other way. You ingratiate yourself with people and now they won't leave you alone. That's when you've got to politely ask for space to think. Every troubleshooter develops their own rhythm over time. Some thrive on constant back-and-forth. Others prefer to work alone. Both approaches can work—but you've got to figure out your balance and stick to it.

At the end of the day, yes, you're the expert walking in. But you also need to be an expert in people as much as equipment. If you're not, you'd better learn—or fake it until you make it. Because no matter how good you are with a scope or a meter, if you don't build rapport and manage boundaries, you'll always be working uphill.

## TOOLS FOR EACH SITVD STEP

### Symptom—Observation Tools

**Your senses**

- **Sound**: Bearings screeching, air leaking, solenoids clicking
- **Sight**: LEDs, position, color changes, smoke (bad day)
- **Smell**: Burning insulation, hot oil, ozone from arcing
- **Touch**: Heat, vibration (careful here)

**Notebook and pen**

Use it for: Write down what you see immediately. In two hours, you won't remember which LED was on.

**Phone camera**

Use it for: Capture the before-state, model numbers, error messages, and wiring before you disconnect it.

**Multimeter—Your Best Friend**

Use it for: This is the one tool you can't troubleshoot without. Spend the money on a good one.

Personal story: My grandfather took me to Shand Electronics in Flint in 1988. Bought me a Fluke 77 for $89—serious money then. That meter has outlasted three companies I've worked for. Still use it today. Lesson: buy once, cry once.

**What you need in a meter:**

- True RMS (for VFD circuits)
- CAT III or IV rating (for industrial use)
- Current measurement (at least 10A)
- Continuity beeper
- Min/Max capture for intermittent signals

**Wiring diagrams**

- Not a physical tool, but essential for isolation. If you don't have prints, you're guessing. Always ask for them first. Take pictures of them. Guard them like gold.

**Pen light**

- Can't fix what you can't see. Panels are dark. Terminal markings are tiny. A good LED penlight costs $20 and saves hours.

## TEST—SIMULATION TOOLS

### Loop Calibrator

- Your second most important tool. Sources 4-20mA to prove the path from field to PLC.

---

Field note: Food plant, flow meter "dead." Operators wanted it replaced. Injected 12mA at the transmitter terminals—HMI showed 50%. Saved them $3,000 and four hours. Problem was a valve upstream that was never opened.

---

### Basic calibrator features needed:

- Source 4-20mA
- Measure 4-20mA
- 24V loop power
- % display for quick math

### Test leads and jumpers

- Good silicone leads (flexible in cold)
- Alligator clips
- Spare fuses for your meter
- Short jumper wires with ferrules

### Oscilloscope (Optional)

- You don't always need one, but when you do, nothing else works.

Field note: Packaging line kept losing encoder counts randomly. Logic looked fine. Wiring tested good. Put the scope on it—saw the signal drop every time the welder three stations down fired. EMI was killing the counts. Rerouted the cable; problem solved. Would never have found it without the scope.

**When you need a scope:**

- Intermittent problems
- Encoder/pulse issues
- Noise hunting
- Communication problems

## VERIFY—CONFIRMATION TOOLS

Independent measurement

- Mechanical pressure gauge
- Thermometer
- Tape measure for level
- Stopwatch for timing

Never trust a single instrument. If the PLC says 50 PSI and your gauge says 0, you know where to look.

Bucket and stopwatch

- Old-school but bulletproof. For flow verification:
  - 5 gallons in 30 seconds = 10 GPM
  - Weight of water = 8.34 lb/gal at 60°F
  - Can't argue with a bucket!

**Document—Recording Tools**

CMMS access

- Get login credentials immediately. Document what you did while it's fresh. Future you will thank present you.
- Tags and labels

Label temporary jumpers

- Tag bad parts before disposal
- Mark calibration dates
- Note configuration changes

## FIELD HACKS AND IMPROVISED TOOLS

Sometimes the best tool is the one you make yourself.

### Cheater Test Lead (Go/No-Go)

- Built one for a Turck junction box system. Took a spare M18 connector, wired in a 24V indicator light. Plug it in—light means powered. No laptop needed, no meter needed. Instant verification. Cost: $15. Time saved: hours.

### Phone with Google Translate

- Had to troubleshoot a machine where the HMI was in English but all the LabVIEW code was commented in Korean. Previous tech walked off the job. I pulled out my phone, turned on Google Translate's camera mode, pointed it at the screen. Real-time translation. Not perfect, but good enough to understand the logic. Got it running in two hours.

### Cold Spray Substitute

- Compressed air duster upside down. Gets cold enough to test thermal intermittents. Not as good as proper freeze spray, but works in a pinch.

## The Back of Your Hand

- For quick temperature checks. Transformers, motors, drives—if it's hotter than it should be, you'll feel it. Just don't touch—hover near it. Old school but effective.

## What Not to Bring

- Your ego
    - Nobody cares about your war stories when the line is down. Save them for break time.
- Every tool you own
    - A cart full of gear makes you look insecure. Bring what you need, get more if required.
- Assumptions
    - "It's always the sensor"—until it's not. Let SITVD guide you, not your hunches.

## Tool Maintenance

Your tools are your livelihood. Treat them right.

Check meter batteries before each job. Verify meter fuses aren't blown. Keep calibrator calibrated. Clean your safety glasses. Update your phone apps. Backup your laptop.

### *The Most Important Tool*

> Your brain. Everything else is just an extension of your troubleshooting method. A tech with a basic meter and solid method beats a tech with every gadget but no system.

# CHAPTER 10
# THE ISOLATION PRINCIPLE

You've got the SITVD framework. You've got your tools. Now you need the roadmap that tells you where to look.

Every signal in an industrial system follows a chain. When something goes wrong, the fault is somewhere in that chain. Your job is to find which link is broken—not by guessing, not by swapping parts at random, but by systematically cutting the problem in half until you've isolated the failure.

This is the Isolation Principle, and it's the difference between a 20-minute fix and a 4-hour wild goose chase.

In a classroom, everything is logical. In the field, it's chaos. You'll have cables running through miles of conduit, sensors buried under layers of grease, and PLC logic written by someone who left the company ten years ago. Without a roadmap, you are just a passenger in that chaos. The Isolation Principle turns you into the navigator.

It forces you to stop looking at the machine as a giant, mysterious "black box" and start seeing it as a sequence of connected events. If the HMI says a tank is empty but you can see it's overflowing, the problem isn't "the machine"—it's a specific break in the communication between the liquid and the screen.

The following pages will break that diagnostic chain down, step by step, showing you exactly how a signal travels from the dirt of the factory floor to the logic of the processor.

## THE DIAGNOSTIC CHAIN

Every signal you troubleshoot follows the same basic path:

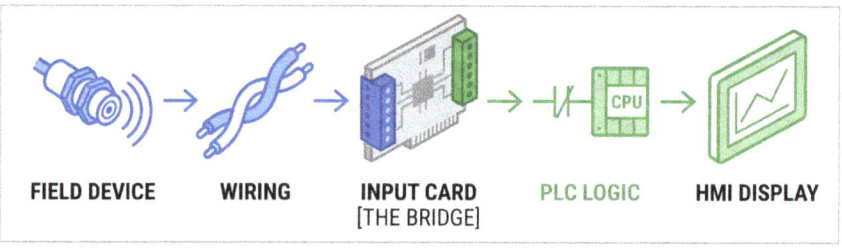

Let's break that down:

| | | |
|---|---|---|
| | **FIELD DEVICE** | **1. Field Device:** The sensor, transmitter, or switch interacting with the physical process. This is the interface—e.g., a photoeye detecting a box, a pressure transmitter measuring PSI, or a solenoid opening a valve. |
| | **WIRING** | **2. Wiring:** The cables, terminals, and junction boxes carrying the signal. It includes both the power wiring for the device and the signal wiring returning to the control system. |
| | **I/O CARD [THE BRIDGE]** | **3. I/O Card [THE BRIDGE]:** The PLC module bridging field signals and processor data. This is where analog becomes digital and 24 VDC becomes a logic "1" or "0". |
| | **PLC LOGIC** | **4. PLC Logic:** The program processing inputs and executing decisions. Faults here often involve scaling errors, wrong addresses, or missing permissives that block outputs. |
| | **HMI DISPLAY** | **5. HMI Display:** The operator interface that reads and displays PLC data. Note: HMI-level scaling or filtering can sometimes make perfectly good data look "bad" to the operator. |

## THE RULE: WORK FROM WHAT YOU CAN SEE

Here's the key insight: **Start where you can observe, then work backward toward what you can't.**

You can see the HMI display. You can usually get online with the PLC and see logic execution and raw values. Getting to the I/O card requires opening a panel. Getting to the field device might require climbing, crawling, or waiting for lockout.

So work backward:

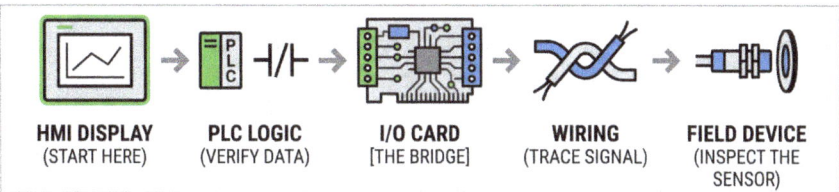

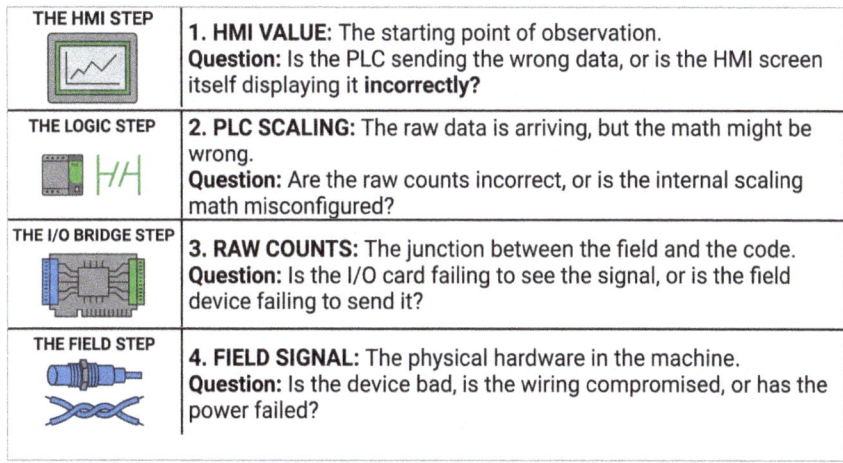

Each question cuts the problem in half. Within 3–4 checks, you've isolated the fault to one link in the chain. This is the Isolation Principle: a repeatable process that replaces guesswork with logic.

We start at the HMI Display not because it's the most likely failure point, but because it's the only place where the entire system's behavior is converted into observable data. Your goal is to find the gap between the physical reality of the machine and the logical reality of the screen.

Before we move into the specific components, let's look at how this roadmap functions during a live failure.

## ISOLATION IN PRACTICE: INPUT SIGNALS

Let's walk through a real example.

**Problem:** HMI shows tank level at 0% but operators say the tank is half full.

**Step 1: Check the PLC Logic**

Is the display configured correctly? Does it show 0% because the PLC is sending 0, or because the HMI tag is broken?

- **Action:** Go online. Find the output rung. Is the rung true? Is the output bit ON?
- **Result:** If the rung is false, a permissive is holding it off. Trace the contacts to find which one is false. That's your problem—it's not the valve, it's the logic.
- **Result:** If the rung is true and the output bit is ON, move to Step 2.

**Step 2: Check the PLC Scaled Value**

The PLC takes raw counts and scales them to engineering units. Is the scaling correct?

Look at the raw counts. For a 4-20 mA signal on a typical Rockwell card, 4 mA ≈ 6,553 counts and 20 mA ≈ 32,767 counts. At 50% level, you should see roughly 19,660 counts.

- If raw counts show ~19,660 but scaled value shows 0%, fix the scaling math.
- If raw counts show ~6,553 (0%), move to Step 3.

**Step 3: Check the I/O Card**

Is the card receiving the correct signal from the field?

- Look at the output LED on the card. Is it lit?
- If the LED is OFF but the logic shows ON, check for a blown fuse on the output card, a tripped output group, or a failed channel.
- If the LED is ON, meter the output terminal. You should see supply voltage (typically 24 VDC).
- If LED is ON but no voltage at terminal, the output transistor or relay is fried. Replace the card.
- If voltage is present at the terminal, move to Step 3.

**Step 4: Check the Wiring**

Is the signal getting from the transmitter to the card?

- With 24 VDC applied, listen for a click or feel for the valve shifting.
- Meter the coil resistance. A good solenoid typically reads 20-200 ohms.
- Infinite ohms = open coil. Near zero = shorted coil.
- If the coil is good but valve doesn't move, it's a mechanical problem—stuck spool, contamination, or broken spring.

**Step 5: Check the Field Device** Is the transmitter working correctly?

- Check that the transmitter has power (loop-powered devices need 24 VDC across the loop).
- Check the local display (if equipped). Does it show the correct level?
- Compare to an independent reference—a sight glass, a dipstick, a portable gauge.
- If the transmitter has power but outputs the wrong signal, it's failed or misconfigured. Replace or recalibrate.

**Total time with isolation method:** 15-20 minutes.

**Time without method (random part swapping):** 2-4 hours, plus wasted parts.

## ISOLATION IN PRACTICE: OUTPUT SIGNALS

Outputs follow the same chain in reverse:

**PLC Logic → I/O Card → Wiring → Field Device**

**Problem:** Solenoid valve won't open. HMI shows "Valve Open" command is active.

**Step 1: Check the PLC Logic** Is the output coil actually energized in the program?

- Go online. Find the output rung. Is the rung true? Is the output bit ON?
- If the rung is false, a permissive is holding it off. Trace the contacts to find which one is false. That's your problem—it's not the valve, it's the logic.
- If the rung is true and the output bit is ON, move to Step 2.

**Step 2: Check the I/O Card** Is the card actually switching the output?

- Look at the output LED on the card. Is it lit?
- If the LED is OFF but the logic shows ON, check for a blown fuse on the output card, a tripped output group, or a failed channel. Some cards have per-channel or per-group fuses.
- If the LED is ON, meter the output terminal. You should see supply voltage (typically 24 VDC) when the output is on.
- If LED is ON but no voltage at terminal, the output transistor or relay is fried. Replace the card.
- If voltage is present at the terminal, move to Step 3.

**Step 3: Check the Wiring** Is voltage getting from the card to the device?

- Meter voltage at the field device terminals. You should see the same voltage you measured at the card.
- If voltage at the card but not at the device, you have an open wire, loose terminal, blown field fuse, or damaged connector.
- Trace the circuit. Check every terminal, junction box, and connector between the card and the device.
- If voltage is present at the device, move to Step 4.

**Step 4: Check the Field Device** Is the device responding to the signal?

- With 24 VDC applied, the solenoid should energize. Listen for a click. Feel for the valve shifting.
- If no response with correct voltage, the solenoid coil is open (burned out) or the valve is mechanically stuck.
- Meter the coil resistance. A good solenoid typically reads 20-200 ohms depending on the model. Infinite ohms = open coil. Near zero = shorted coil.
- If the coil is good but valve doesn't move, it's a mechanical problem—stuck spool, contamination, broken spring.

## THE HALF-SPLIT METHOD

When you hit a troubleshooting roadblock, use the **Half-Split Method** to instantly simplify the problem. Instead of checking every connection in a sequence, this method divides the circuit, systematically eliminating entire sections. By focusing on a logical midpoint, like a junction box or terminal strip, you can determine if the failure is "upstream" (closer to the source) or "downstream" (closer to the destination), often isolating the exact fault with just two or three strategic measurements.

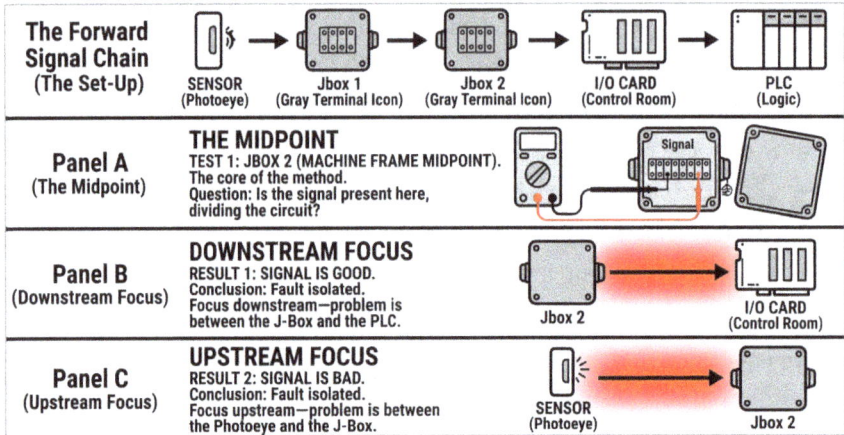

## COMMON ISOLATION MISTAKES

### Mistake 1: Starting at the wrong end

Don't climb a ladder to check a sensor when you haven't even looked at the PLC yet. Always start where it's easy—usually the HMI and PLC—and work outward only as needed.

### Mistake 2: Skipping links in the chain

If you check the HMI and then jump straight to the field device, you'll miss problems in the PLC scaling, the I/O card, and the wiring. Every link matters.

### Mistake 3: Assuming instead of testing

"The wiring's fine, I just checked it last week." Did you? Prove it. Meter it now. Things change. Vibration loosens terminals. Mice chew cables. Last week doesn't matter.

## Mistake 4: Swapping parts without isolating

You replaced the sensor, but the problem was actually a loose terminal at the card. Now you've wasted a sensor and still have the same fault—and you've lost confidence because "the new sensor didn't fix it."

## Mistake 5: Forgetting the power supply

Every field device needs power. A 4-20 mA transmitter won't output anything if there's no 24 VDC loop power. A proximity sensor won't detect anything if its supply is missing. Always verify power first.

# THE JUMPER TEST: YOUR SECRET WEAPON

When you're not sure if the problem is the field device or the wiring/card, use a jumper test.

**For discrete inputs:**

1. Disconnect the sensor
2. Jumper the input to the active state (+24V for PNP, 0V for NPN)
3. Does the PLC see the input?

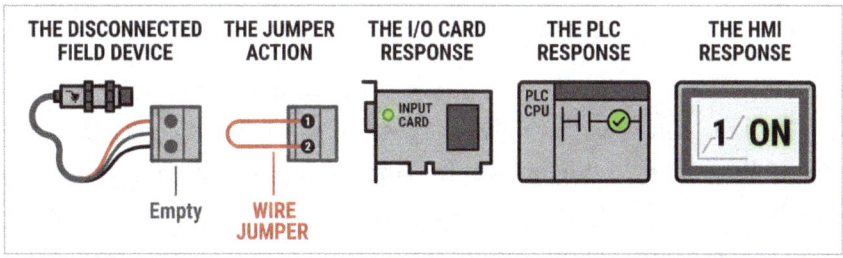

If yes → Problem is the sensor or its wiring If no → Problem is the card or PLC configuration

**For analog inputs:**

1. Disconnect the transmitter
2. Inject a known signal with a loop calibrator (e.g., 12 mA = 50%)
3. Does the PLC show 50%?

If yes → Problem is the transmitter If no → Problem is the card, wiring, or scaling

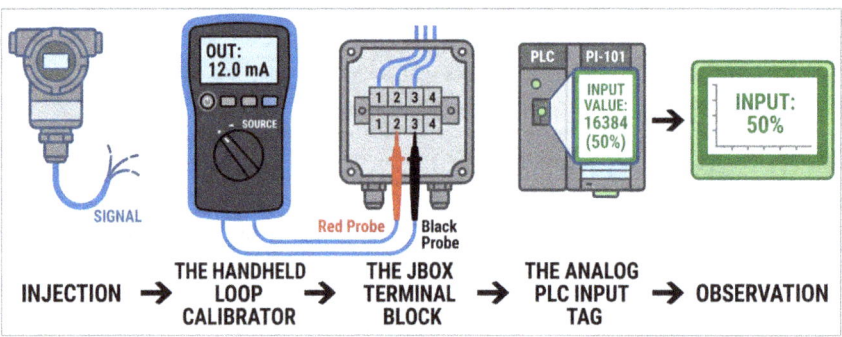

The jumper test isolates the entire field side from the control system in one simple check. It's fast, cheap, and decisive.

However, a word of caution: beware of the *False Positive*. When you inject a signal at the junction box and see 50% on the HMI, you have proven that the wiring from the box to the PLC is good. You haven't yet proven the wiring from the sensor to the box.

This is why the location of your jumper matters. If you jumper at the I/O card and the signal works, you've narrowed the problem down to the "field." But that field still includes five hundred feet of conduit, three junction boxes, and a connector at the sensor. Always work your jumper tests from the control room outward to the device to systematically "clear" sections of the path.

Once the jumper proves the system can see a signal, you can stop fighting the software and the wiring. You have successfully isolated the fault to the physical device or its immediate connection.

## DOCUMENT YOUR PATH

As you isolate a fault, write down what you checked and what you found:

- HMI: Shows 0%
- PLC scaled value: 0%
- PLC raw counts: 6,553 (equals 4 mA)
- Current at card: 4 mA
- Current at transmitter: 4 mA
- Transmitter local display: Shows 50%

That trace tells you the transmitter knows the right value but is outputting the wrong signal. The transmitter's output circuit is bad, or it's configured wrong. You've isolated the problem without swapping a single part.

If you don't document your path, you'll forget what you checked. You'll repeat tests. You'll waste time. Write it down.

## ISOLATION IS A DISCIPLINE

The Isolation Principle isn't complicated. It's just disciplined.

Every fault is somewhere in the chain. Your job is to find it methodically, not randomly. Start where you can see. Work toward what you can't. Cut the problem in half with each test. Prove each link before moving on.

---

⚡ **Shop-Floor Wisdom:** The fastest troubleshooters aren't the ones who know the most—they're the ones who waste the least time looking in the wrong place.

---

# PART III
# KNOW YOUR MACHINE

# CHAPTER 11
# DISCRETE INPUTS

Discrete inputs are the simplest signals you'll work with—either ON or OFF, 1 or 0, TRUE or FALSE. A limit switch is closed or open. A proximity sensor sees metal or it doesn't. A push button is pressed or released.

But don't let the simplicity fool you. Discrete inputs are simple in concept but surprisingly easy to screw up in practice.

## INPUT TYPES & WIRING

### 2-Wire Sensors (Dry Contact)

The simplest type. Two wires, acting like a switch—either open or closed. When closed, current flows. When open, it doesn't.

Examples:

- Limit switches, Pressure switches, Float switches
- Reed switches, Push buttons

### The Reliability of Simplicity

In a world of high-tech laser sensors and smart transmitters, the humble dry contact is still your best friend. Why? Because they are "passive." They don't have processors that can freeze up or firmware that needs updating. If the mechanical arm of a limit switch moves, the circuit closes.

When you are troubleshooting these, remember that the most common failure isn't the switch—it's the environment.

- **Mechanical Wear:** Is the actuator arm actually hitting the switch?
- **Contamination:** Has grease or coolant seeped into the housing and insulated the contacts?
- **Loose Terminals:** Vibrating machines love to back out terminal screws over time.

Before you even pull out your meter, look for the "Mechanical 1 and 0." If the switch isn't physically clicking, the wiring diagram on the next page won't help you.

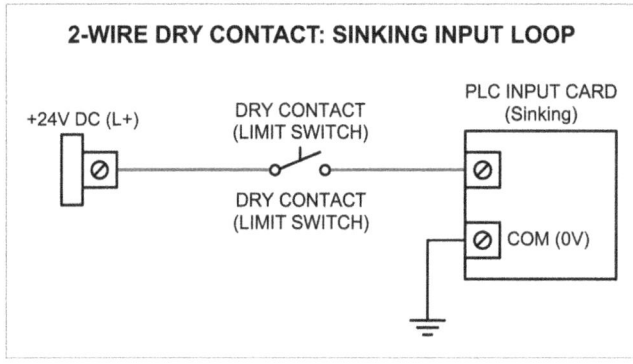

*Illustration 2-wire dry contact: Sinking Input Loop*

---

## 3-Wire Sensors (Solid State)

Most modern proximity sensors and photoelectric sensors are 3-wire: +24V (brown), 0V/common (blue), and signal output (black or white).

Every 3-wire sensor needs all three wires properly connected: +24V, 0V, and signal. If the common (0V) wire is loose, broken, or not landed, the signal wire floats. A floating input picks up electrical noise and reads erratically—sometimes ON, sometimes OFF, never reliable.

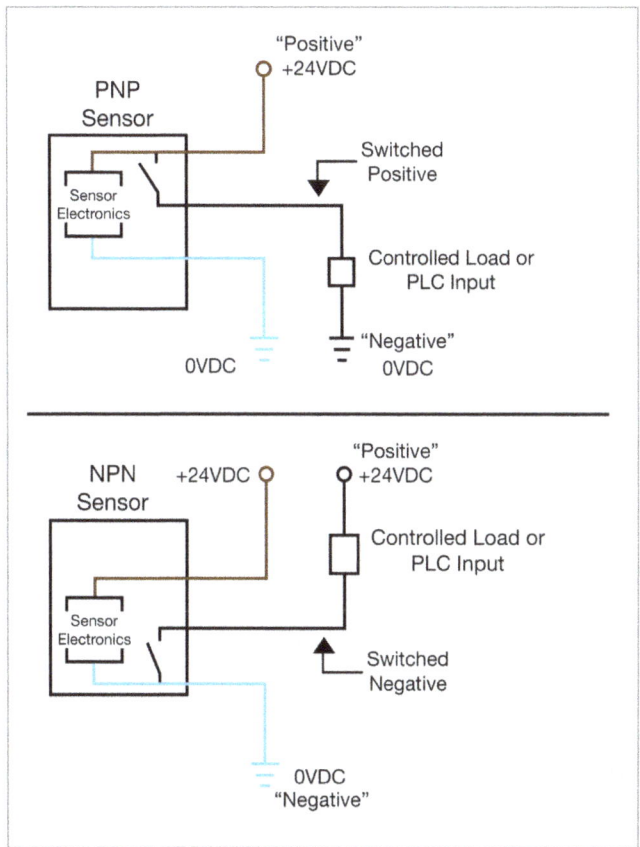

*Illustration 3-wire solid state sensors*

## WHY THIS MATTERS:

If you connect a PNP sensor to a card expecting NPN, or vice versa, the input will never work correctly. The card's internal circuit provides +24V. When an NPN sensor activates and pulls the signal wire to 0V, current flows from the card's internal supply through the input circuit and out through the sensor to ground. The card sees this current flow as "ON." The LED on the sensor will light up, but the PLC input bit will stay OFF (or always ON, depending on the wiring). This is the #1 cause of "the new sensor doesn't work" calls.

## INPUT CARD TYPES

Modern PLC input cards are designed to detect the presence or absence of a voltage signal. In a DC environment, they are categorized by how the current flows between the field device and the card.

| Card Type | Electrical Logic | Typical Application |
|---|---|---|
| Sinking Input Card | Expects a HIGH (+24V) signal. Current flows into the card to ground. | Standard for PNP sensors (most common in North America). |
| Sourcing Input Card | Expects a LOW (0V) signal. Current flows out of the card to the device. | Standard for NPN sensors. |
| Universal / Configurable | Can be wired or programmed for either sinking or sourcing logic. | Varies; check the "per-channel" or "per-group" documentation. |

*Table Modern PLC Input Card Types*

Many modern modules are **Universal PLC Input Cards**, designed to support both Sinking and Sourcing configurations depending on how you wire the Common terminal. By tying the Common to +24V, you set the card to **Sinking Mode**, which allows NPN sensors to complete the circuit by pulling the input to ground.

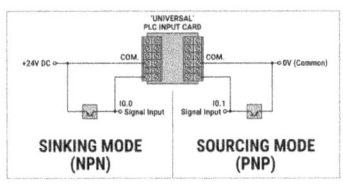

*Illustration Universal PLC Input Card*

If you tie the Common to 0V, the card switches to **Sourcing Mode**, enabling PNP sensors to trigger the input by sending a positive 24V signal. This flexibility allows a single spare part to support almost any discrete input configuration in your facility.

## A NOTE ON AC DISCRETE INPUTS

Most modern discrete I/O runs on 24 VDC. If you're in an older plant—particularly pre-1990s North American equipment—you may encounter 120 VAC input cards. The troubleshooting logic is identical: verify voltage at the terminal, verify the bit state online, follow the chain. The key differences are the threshold voltages (a 120 VAC card typically needs >80 VAC to register ON and <20 VAC to register OFF) and the leakage current behavior, which is higher and more variable than solid-state DC sensors. A standard DC meter will show RMS voltage correctly on an AC input card, but never work on AC circuits without confirming your meter's CAT rating first.

## COMMON FAILURE MODES

Discrete input failures usually fall into one of three categories: wiring problems, configuration errors, or sensor and environmental issues. While a "no signal" fault might seem simple, the root cause can hide anywhere from a loose terminal in a junction box to a software debounce timer set too high.

To troubleshoot effectively, you must categorize the symptom before you reach for your tools. A signal that is **intermittent** usually points to physical wiring fatigue or environmental interference. A signal that is **completely dead** often suggests a blown fuse or a total component failure. A signal that **"ghosts" or flickers** without a target is almost always a configuration or polarity mismatch.

The following table maps these common symptoms to their likely causes and provides a "First Test" for each. Use this as your rapid-response guide to narrow down the problem before beginning a deep-dive investigation.

## DISCRETE INPUT FAILURE MODES

| Failure Mode | Symptom | Likely Cause | First Test |
|---|---|---|---|
| **WIRING ISSUES** | | | |
| FLOATING COMMON | Flickering or Erratic | Loose wire | Meter 0V at both ends |
| DAMAGED WIRE | Intermittent | Fatigued wire | Wiggle test |
| WRONG POLARITY | Input never works/stuck | Terminal wiring reversed | Check documentation |
| BLOWN INPUT FUSE | Group of Dad | Short / overcurrent | Check fuse |
| **CONFIGURATION ERRORS** | | | |
| PNP/NPN MISMATCH | LED; PLC never sees it | Sensor mismatch | Check sensor label |
| FILTER TOO LONG | Misses fast events | Debounce high | Reduce filter time |
| RELAY FAILED | Input doesn't track | Interposing relay coil or bad contac | Measure 24VDCC |
| **SENSOR/ENVIORNMENTAL** | | | |
| SENSOR DEAD | No LED / Stuck | Transmitter failed | Jumper test |
| ENVIRONMENTAL | Misses Target / False Trigger | Dirty Lens | Clean sensor face |
| CONTACT WEAR | Stuck ON, OFF, Intermittent | Welded or pitted | Check continuity |

*Table of Common Failure Modes*

## THE VOLTAGE LEAK TRAP

⚡ **Shop-Floor Wisdom:** "If the LED is on but the PLC is blind, trust your meter, not the light."

Solid-state sensors do not always switch cleanly to absolute zero. When a PNP sensor is "off," it typically leaks **1–3V** on the signal line. While usually harmless, a failing or partially-damaged sensor can leak **8–15V** while appearing normal. This leaked voltage is often enough to forward-bias the input circuit, causing the PLC to register a logic **HIGH (1)** even when no target is present.

### How to test for it:

| Test Scenario | Good Sensor | Leaky Sensor Trap |
|---|---|---|
| Meter at Terminal (Target Absent) | 0–2V | 8–15V |
| Meter at Terminal (Target Present) | 20–24V | 20–24V |
| PLC Input Bit (Target Absent) | 0 (OFF) | 1 (ON) |
| PLC Input Bit (Target Present) | 1 (ON) | 1 (ON) |

*Table Voltage Leak Trap*

## What the Numbers Mean

See voltage between 5V and 15V? That's leakage from a solid-state sensor not fully switching off. It's usually harmless unless it's high enough to forward-bias the input and register as HIGH when it shouldn't.

The fix: replace the sensor, add a bleeder resistor (1kΩ to 4.7kΩ across the input), or use a card rated for solid-state sensors.

**Quick Test:** Disconnect the sensor at the card. Input drops to zero = sensor is fine, check your logic. Input stays HIGH = card problem or wiring short.

## TROUBLESHOOTING DISCRETE INPUTS

Sometimes the sensor is working perfectly—it's just not seeing what you think it should see. To find out where the disconnect is happening, we return to the SITVD chain.

You previously saw this chain as a horizontal roadmap, but in practice, troubleshooting is a **top-down checklist.** We start at the top with the symptom (the HMI) and work our way down to the physical device. By treating the SITVD method as a vertical stack, you can physically "check off" each level of the system until the fault is found.

Follow the chain: **Sensor → Wiring → Card → PLC Logic**

| Step | Signal Chain | Action |
|---|---|---|
| Symptom | PLC CPU / HMI | Observe the error: "Input I:0/1 not responding." |
| Isolate | Wiring / Terminal / Card | Use the Jumper Test to see if the PLC card is alive. |
| Test | The Sensor / Switch | Actuate the device and meter the signal at the source. |
| Verify | The Entire Loop | Confirm the bit toggles in logic when the field moves. |
| Document | Work Order / Schematic | Mark the failed sensor for replacement and log the fix. |

## SPECIAL CASES

### Input Filters / Debounce

Many PLC input cards have configurable input filters that delay the response to prevent false triggers from noise. If the filter is set too long (e.g., 20ms) and your sensor is seeing fast events, the input may appear to miss pulses or respond slowly.

Check the card configuration and reduce the filter time if needed—but not so low that you pick up noise.

Most input cards default to 10ms filtering, which works for 95% of applications. Only adjust it if you have a specific reason—either your process is genuinely faster than 10ms, or you're getting false triggers that shouldn't be there. When in doubt, leave it alone and look elsewhere for your problem.

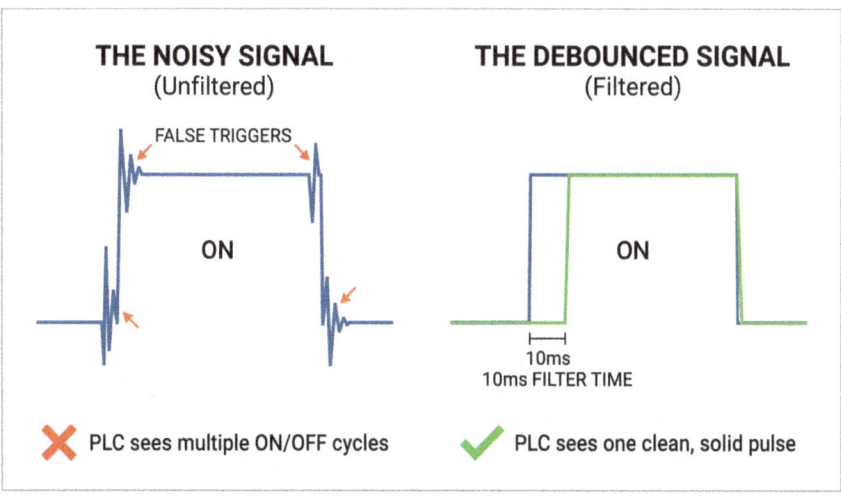

## Interposing Relays

**Interposing Relays & Common Failures** Interposing relays bridge the gap between field devices and PLC inputs, providing **electrical isolation**, **voltage conversion**, or **contact multiplication**. While these relays add a layer of protection, they also introduce new failure points. Coils can fail open from moisture or age, and contacts eventually weld shut or pit from millions of cycles. In high-vibration environments, terminal screws can back out, leading to intermittent signals. Importantly, a relay may physically "click" even if the internal electrical contacts have failed to make a clean connection.

Troubleshooting Strategy

Verify voltage at the coil (**A1/A2**) when the device is active. If the coil is energized but the relay doesn't click, swap it. If it clicks but the PLC is blind, meter the output contacts for continuity. If the contact is closed but the signal isn't reaching the PLC, the fault is in the downstream wiring or the card itself.

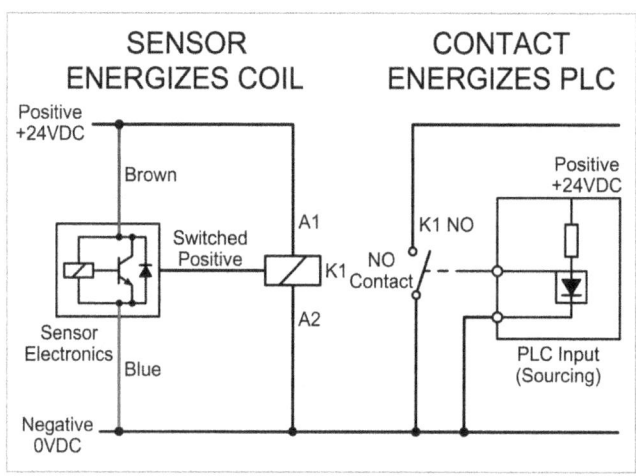

# BEFORE YOU REPLACE IT—DISCRETE INPUT CHECKLIST

Run through this before pulling any sensor or card. Takes under five minutes.

- ☐ **Input card channel LED**—Before going online, look at the card itself. Most modern input cards have a green channel LED per point. If the LED tracks the sensor (lights when target is present), the signal is reaching the card. If the LED is correct but the PLC bit is wrong, suspect card configuration or slot assignment, not the field device.

- ☐ **Sensor LED**—Does it change state when target is presented/removed?

- ☐ **Sensor type vs. card type**—PNP sensor → sinking card? NPN sensor → sourcing card?

- ☐ **Common wire**—Is the 0V (blue) wire fully landed at the sensor and at the PLC terminal?

- ☐ **Signal wire voltage**—Meter the black wire at the PLC terminal. Active = 20–24V (PNP). Inactive = 0–2V.

- ☐ **Jumper test**—Jumper the input terminal to +24V. Does the bit go ON? If yes: problem is field side. If no: problem is card or configuration.

- ☐ **Wiggle test**—Go online and flex the cable. Does the bit flicker? If yes: intermittent wire, usually at the connector or a bend point.

- ☐ **Input filter setting**—Is the card filter time set longer than your event duration? Fast events can be "missed" by a slow filter.

## HIGH-SPEED INPUTS

Some applications—encoders, pulse-counting, registration marks—require dedicated high-speed input channels that bypass the standard filter. If you're troubleshooting a system that counts pulses or tracks position, verify the input is assigned to a high-speed channel. Standard input channels will miss fast pulses regardless of filter setting. High-speed input troubleshooting is covered in Book 2: PLC Troubleshooting.

## REAL-WORLD EXAMPLE

**Problem:** Part-in-place sensor on a weld fixture never shows TRUE. Robot waits forever for the signal.

**Symptom:** HMI shows "Waiting for Part in Place." Operators say the part is definitely seated.

**Troubleshooting:**

1. **Check sensor LED:** LED is ON when part is in nest. Sensor sees the target.
2. **Meter at PLC terminal:** With part in place, voltage at input terminal is 0V. Without part, voltage is also 0V. That's wrong—should see +24V when sensor is active if it's PNP.
3. **Check sensor wiring:** Brown wire to +24V—good. Blue wire to 0V—good. Black wire (signal) to input terminal—good. Wait—the blue wire at the sensor is loose. Pushed into the connector but not actually in the terminal.
4. **Fix:** Re-landed the blue (common) wire at the sensor connector. Voltage at input terminal now shows +24V when part is present. PLC input bit goes TRUE. Robot continues cycle.

**Total time:** 12 minutes.

**Root cause:** Floating common from loose wire at sensor connector. *Sensor appeared to work (LED lit) because it had enough leakage current to power the LED, but the output couldn't switch properly without a solid common reference.*

---

### ⚠ SAFETY-RATED I/O—DO NOT APPLY STANDARD TECHNIQUES

If the circuit you are troubleshooting is connected to a safety relay, safety PLC, or is labeled as a safety-rated input or output—E-stop loops, light curtains, two-hand controls, safety gate switches—stop. Standard troubleshooting techniques (output forces, input jumpers, bypassing interlocks) can disable the safety function and create an immediate hazard. Safety-rated circuits must be troubleshot following the machine's safety documentation and applicable standards (ISO 13849, IEC 62061). When in doubt, consult the OEM documentation or a qualified functional safety engineer before proceeding.

### DECISION TREE REFERENCE

See **Appendix B: Discrete Input Troubleshooting** for step-by-step flowcharts you can photocopy and laminate.

# CHAPTER 12
# DISCRETE OUTPUTS

If inputs tell the PLC what is happening, outputs are how it makes things happen—opening valves, starting motors, or energizing solenoids. When an output fails, production stops. While these signals are binary, their failure modes are varied: welded relay contacts, blown fuses, or transistors destroyed by inductive loads without suppression.

## THE OUTPUT CHAIN

Understanding the circuit is the key to rapid isolation. Power originates at the supply, passes through the card, energizes the load, and returns via the common or neutral path. The card only acts as a switch; it does not create power. A dead supply or an open return path will stop the output regardless of the PLC's status.

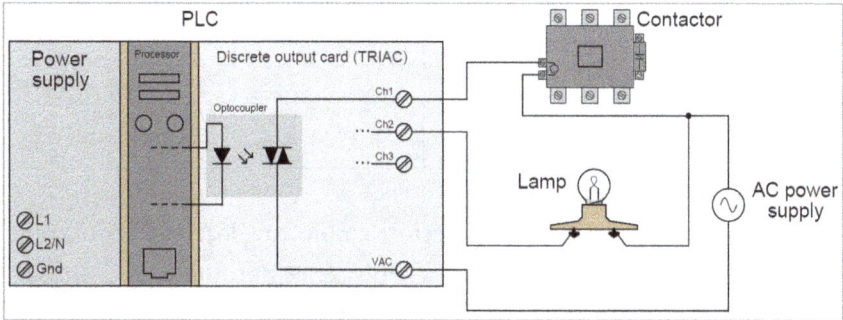

## What You'll Learn

This chapter covers how outputs function, why they fail, and how to isolate problems quickly. You will learn the critical differences between relay, transistor, and triac outputs, and why matching the wrong load type leads to immediate hardware failure. Finally, we will trace the power path from PLC logic to the field device using a systematic troubleshooting sequence.

## OUTPUT TYPES

Not all output cards are the same. The type of output determines what loads it can drive, how fast it can switch, and how it fails.

| Output Type | Best For... | Advantage | Typical Failure |
|---|---|---|---|
| Relay | AC/DC, High Current | Flexible, Dry Contact | Welded Contacts (Stuck ON) |
| Transistor | 24VDC, Fast Switching | Unlimited Cycle Life | Shorted Transistor (Stuck ON) |
| Triac | 120VAC, Inductive | No Mechanical Wear | Latch ON (if used on DC) |

⚠ **Critical:** Know your card type before connecting loads. Relay outputs are flexible. Transistor outputs are DC only. Triac outputs are AC only. Mixing them up causes failures and confusion.

If someone connects a DC load to a triac output, the triac may latch ON and never release. The output is commanded OFF, the LED is OFF, but the load stays energized. Check the load voltage—if it's DC and the card is triac, that's your problem. I've seen this confuse techs for hours. Output cards switch power to a load. The basic circuit is:

**Power Supply → Output Card → Load → Return Path (Common/Neutral)**

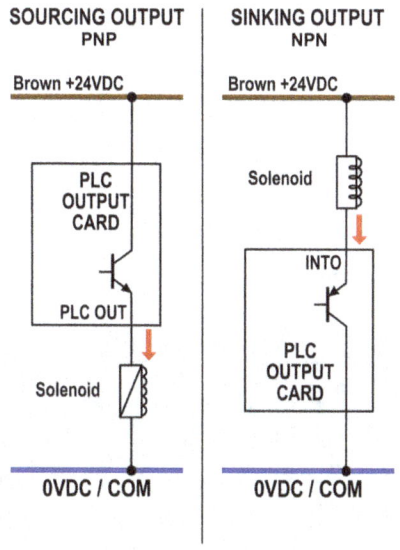

*Illustration Sourcing Output PNP and Sinking Output NPN circuits*

**Important:** Many output cards have grouped commons—a single common terminal for 4, 8, or 16 outputs. Check the card documentation. If you mix AC and DC loads on a card with grouped commons, you can create dangerous short circuits.

## FUSE SIZING FOR GROUPED OUTPUTS

Each output group typically shares a single fused common. When that fuse blows, all outputs in the group go dead simultaneously—which looks exactly like a failed output card on first inspection. Before condemning a card, measure voltage at the common terminal for each group. No voltage at the common = blown fuse, not a bad card. Check the card documentation for fuse type and rating. Never upsize a fuse to "solve" repeated blows—find the overloaded or short-circuited load first.

## LOAD CONSIDERATIONS

Not all loads are the same. Understanding what you're driving helps you predict and prevent failures.

| Load Type | Risk | Source | Prevention |
|---|---|---|---|
| Resistive (Heaters/Lamps) | LOW | Steady current; no spikes. | None required. |
| Capacitive (LED Drivers) | LOW | Initial capacitor charging current. | Usually handled by card design. |
| Lamp (Incandescent) | MEDIUM | Cold filament has 10–15x inrush current. | Derate output or switch to LED. |
| Inductive (Solenoids/Relays) | CRITICAL | Collapsing magnetic field creates high-voltage back-EMF. | Flyback Diode (DC) or RC Snubber (AC). |

### Matching Output to Load

Before you wire anything, verify the output card can handle what you're asking it to drive. Check three specs: continuous current rating, surge current rating, and switching frequency. A card rated for 2A continuous might handle a 100mA pilot light forever, but a 1.5A solenoid with 10A inrush will kill it in a week.

The load table above tells you what to watch for. Resistive loads are easy—if it draws 0.5A and the card is rated 2A, you're fine. Inductive loads are harder because the inrush current isn't on the nameplate. Measure it, or use the manufacturer's data, or assume it's 6-10 times the running current and size accordingly.

**Field Reality:** Most output card failures come from inrush current on inductive loads, not steady-state overload. A contactor coil rated at 0.2A might pull 1.5A for the first 50ms. Do that a million times and the output transistor degrades. Add suppression and it lasts forever.

## LOAD SUPRESSION

Whenever you turn off an inductive load—like a solenoid, relay coil, or motor starter—the magnetic field supporting that device collapses instantly. This collapse creates a massive voltage spike, often five to ten times the supply voltage, which travels backward toward your PLC. Without suppression, this "inductive kick" will eventually arc across relay contacts (welding them shut) or punch a hole through the silicon of a transistor output.

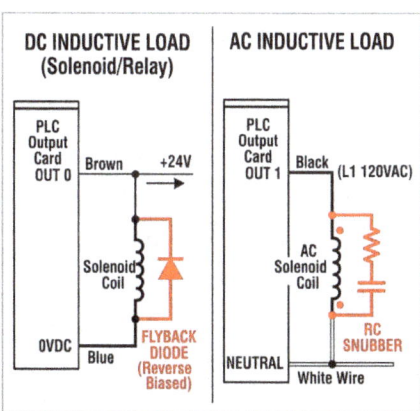

Suppression is the practice of giving that high-voltage spike a safe path to dissipate before it reaches your expensive hardware.

### Installation Best Practices

Mount suppression components as close to the load as possible, not at the PLC card. If you install a flyback diode at the output card and run 50 feet of cable to the solenoid, the voltage spike still travels down the cable and hits the card before the diode can clamp it. Put the diode across the solenoid terminals where it belongs.

For AC snubbers, wire them directly across the contactor coil terminals. Keep the leads short—less than 6 inches if possible. Long leads add inductance that defeats the purpose of the snubber.

### When Suppression Fails

If outputs keep failing even with suppression in place, the suppression isn't working. Check the flyback diode polarity—install it backwards and it does nothing. Check for loose connections at the snubber. Measure the voltage spike with an oscilloscope if you have one; if the spike is still above 100V on a 24VDC system, your suppression is inadequate.

Some loads need more aggressive suppression than others. A small solenoid valve might be fine with a standard 1N4004 diode. A large contactor might need a 3A diode or even a MOV (metal oxide varistor) in parallel with the flyback diode for extra clamping.

### The Bottom Line

Every inductive load needs suppression. No exceptions. It's not optional, it's not something you add "if you have problems." It's part of the correct installation. Add it up front and your output cards will last for years. Skip it and you'll be replacing cards every few months.

---

## COMMON FAILURE MODES

| Failure Mode | Primary Symptom | Root Cause | SITVD Focus |
|---|---|---|---|
| **Welded Relay** | Load stays ON; LED is OFF. | Inrush current fused contacts. | T (Test across contact) |
| **Shorted Transistor** | Load stays ON; LED is OFF. | Inductive spike/ Overcurrent. | V (Verify suppression) |
| **Blown Fuse** | LED is ON; No voltage at terminal. | Shorted load or wiring. | I (Check card status) |
| **Open Coil (Load)** | Voltage at load; No movement. | Burned out solenoid/ contactor. | V (Meter coil Ω) |
| **Wiring Fault** | Voltage at card; 0V at load. | Loose terminal/ Broken wire. | V (Trace the path) |
| **Logic Interlock** | Output never turns ON; LED OFF. | False permissive/ Safety active. | S (Check the HMI) |

While the matrix above identifies the most likely physical culprits, remember that a system is only as reliable as its logic. Before you break out the multimeter to chase a suspected hardware failure, you must first verify that the software isn't the actual bottleneck. The following sections cover the two most common 'logic ghosts': manual overrides and misconfigured interlocks.

## FORCES AND OVERRIDES

PLC programming tools allow technicians to "force" outputs ON or OFF, overriding the logic. Forces are useful for testing, but dangerous if left in place.

## Before blaming hardware:

| Look For... | Why It Matters | Hardware Impact |
|---|---|---|
| Force Table | A non-empty table is your first clue. | Overrides physical card status. |
| The "F" Icon | Software tags marked with F or ⚡. | Output is "Frozen" and deaf to logic. |
| Logic Permissives | Open rungs in the PLC program. | Signal is blocked before reaching the card. |
| Safety Interlocks | Active E-Stops or Light Curtains. | Hardware power is physically removed. |

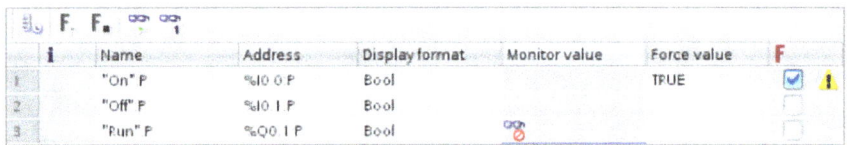

*Image courtesy of Siemens AG*

## The "Ticking Time Bomb" (Safety & Liability)

⚠ **SAFETY WARNING:** Forces are a significant liability. In many systems, they can bypass internal logic permissives and safety interlocks. A force left in the table is a "ticking time bomb" for the next shift; never consider a job finished until the **Force Enable** light on the PLC processor is **OFF**.

## TROUBLESHOOTING WORKFLOW

Follow the chain: **PLC Logic → Output Card → Wiring → Load**

| Step | Signal Chain | Action |
|---|---|---|
| Symptom | The Load | Device isn't moving, but HMI says it should be. |
| Isolate | PLC Output LED | Is the PLC actually trying to turn it on? Check the LED. |
| Test | Output Terminal | Meter voltage at the card. Supply voltage present? |
| Verify | Field Wiring/Load | Meter at the load. Voltage present but no movement? |
| Document | Schematic/Log | Document the fix (e.g., 'Added suppression diode'). |

## THE METER TRAP: GHOST VOLTAGES

Digital multimeters have very high internal resistance (impedance). While this is great for accuracy, it often picks up "ghost" or "leakage" voltage on PLC output cards—especially those using solid-state transistors or Triacs. Your meter may show 24VDC or 120VAC, leading you to believe the card is good, yet the load never moves.

**The Rule of Thumb:** If you suspect a ghost voltage, perform the **Load Test** (see next page). By placing a physical test load on the terminal, you force the circuit to do real work. If the voltage vanishes under load, the output component has failed "open" or is leaking.

## THE LOAD TEST: PROVING THE OUTPUT

Use a known test load to prove the output card is good.

| Output Type | Recommended Test Load | Why it Works |
|---|---|---|
| 24VDC | 24V Indicator Lamp (~100mA) | Draws enough current to prove transistor is switching. |
| 24VDC | 1kΩ Resistor (~24mA) | Minimum "clean" load to verify voltage drop. |
| 120VAC | 120V Trouble Light | High visibility; confirms Triac or Relay is firing. |
| 120VAC | Small 120V Relay | Proves the output can handle an inductive "kick." |
| **PROCEDURE** | | |
| 1 | Disconnect the field wiring from the output terminal | |
| 2 | Connect test load between the output terminal and return | |
| 3 | Set PLC output to **ON** | |
| 4 | Is test load energized? | |
| **YES** | Output card is good. Problem is field wiring or original load. | |
| **NO** | Output channel is bad. Check fuses first, then replace card. | |

# REAL-WORLD EXAMPLES

## Example 1: The Phantom Solenoid

**Problem:** Divert gate solenoid never energizes. Parts pile up at the divert station.

**The Process:** Following the **S.I.T.V.D.** workflow, we confirmed the PLC logic was TRUE and measured 24VDC at the output card terminal. However, a meter check at the solenoid showed 0VDC, indicating a break in the field wiring.

**Finding:** A junction box terminal screw had backed out due to vibration. Tightening the terminal restored the signal.

**Time:** 15 minutes

## Example 2: The Cooked Transistor

**Problem:** A new machine's pneumatic cylinder failed to extend on the first day of production.

**Finding:** The output transistor was "open" (dead). We traced the field wiring and discovered the solenoid coil lacked a suppression diode. The inductive spike from the first few cycles killed the transistor.

**Fix:** Replaced the output card and added a flyback diode to the solenoid.

**Time:** 45 minutes (plus adding diodes to 8 solenoids)

**Lesson:** Always check for inductive load suppression on new installations. It's cheap insurance.

## PREVENTIVE MEASURES

Effective troubleshooting is only half the battle; the ultimate goal is to eliminate the root causes of failure before they stop production. While industrial environments are inherently harsh, most discrete output failures are preventable through disciplined installation and maintenance standards. By implementing the following preventive measures, you can significantly reduce unexpected downtime and extend the lifecycle of your automation hardware.

| Action Item | Technical Requirement | Prevention |
|---|---|---|
| Add Suppression | Diodes (DC), Snubbers/MOVs (AC) | Prevents inductive "kick" from killing transistors. |
| Monitor Current | Use interposing relays for >2A loads | Protects delicate card circuitry from overcurrent. |
| Inventory Spares | Fuses, relay modules, and 1 spare card | Eliminates "2 AM downtime" while waiting for shipping. |
| Audit Forces | Log and clear all software overrides | Prevents "ticking time bombs" for the next shift. |
| Check Torque | Periodic terminal tightening/thermal scans | Prevents high-resistance heat and fire hazards. |

### The Cost of Prevention vs. Repair

Every item in the table above takes time and costs money up front. Suppression components, interposing relays, spare cards, software audits, torque checks—none of it is free. But compare that cost to a single unplanned shutdown at 2 AM: lost production, emergency callouts, expedited parts shipping, and the ripple effects on downstream processes.

Preventive measures pay for themselves the first time they prevent a failure. The second time, they're pure profit. Install suppression on every inductive load today, and you won't be replacing output cards every six months. Audit your forces quarterly, and you won't troubleshoot phantom failures caused by overrides someone left active three years ago.

---

## DECISION TREE REFERENCE

See **Appendix B: Discrete Output Troubleshooting** for a step-by-step flowchart you can photocopy and laminate.

⚡ **Shop-Floor Wisdom:** Trust your meter more than the LED. The PLC may think it turned something on, but the plant only cares if the device actually moved.

💡 **Pro Tip:** Don't just swap output cards. If the load is bad or unsuppressed, the new card will fail too. Find the root cause before replacing hardware.

💡 **Pro Tip**: Relay output contacts can test good on a cold ohmmeter (near 0Ω) but fail under load. Arcing during switching erodes the contact surface over time. A relay that reads fine on the bench but won't hold a 24 VDC solenoid coil in service has arc-eroded contacts. The load test—not the ohmmeter—is the definitive check. Relay output cards have rated electrical lives (number of switching operations under load); in high-cycle applications, track switch counts if your PLC platform exposes them.

# CHAPTER 13
# ANALOG LOOPS

Discrete signals are binary—**ON** or **OFF**. Analog signals are proportional; they quantify the real-world process. Whether it is Pressure (PSI), Temperature (°F), or Flow (GPM), the analog loop is the primary data link between the field and the controller.

When any link in the chain—power, wiring, transmitter, input card, or scaling—fails, the resulting data is corrupted. To a PLC, "Bad Data" is often indistinguishable from "Truth" without systematic troubleshooting. This chapter provides the framework for verifying loop integrity from the sensor to the HMI.

## THE 4-20 MA STANDARD

The 4-20 mA current loop is the dominant standard in industrial instrumentation. Understanding why it works the way it does helps you troubleshoot it.

### Why Current Instead of Voltage?

With a 0-20 mA system, you can't tell the difference between "zero flow" and "broken transmitter." Both read 0 mA. With 4-20 mA, a reading of 0 mA always means something is wrong—the loop is open, the transmitter is dead, or there's no power.

| Factor | Voltage (0–10V) | Current (4–20 mA) |
|---|---|---|
| Signal Loss | *IR* Drop: DC attenuation over distance. | Consistent: *I* is identical in series. |
| Noise | High-*Z*: Susceptible to EMI/RFI. | Low-*Z*: High noise immunity. |
| Integrity | Ambiguous: 0V = 0% or Broken Wire. | "Live Zero": 4 mA = 0%. 0 mA = Fault. |
| Max Run | Short (Panel to Panel). | Long (Field to Panel). |

Current loops are robust, but they require a closed circuit. If the path is broken, the intelligence is lost.

### Live Zero vs. Dead Zero

The 4-20 mA standard uses 4 mA for 0% instead of 0 mA. This is intentional. When you measure 0 mA, you immediately know something is wrong—open wire, dead transmitter, no power. You don't have to wonder if the process is actually at zero or if the loop is broken. With 0-10V systems, 0V could mean either "process at zero" or "wire broken." That ambiguity wastes time. With 4-20 mA, 0 mA always means failure.

## LOOP ARCHITECTURES

| Architecture | Power Source | Signal Path | Best For... |
|---|---|---|---|
| 2-Wire | Loop (PLC) | Shared Pair | Standard field instruments. |
| 3-Wire | External | Shared Common | High-power sensors / shared common. |
| 4-Wire | External | Fully Isolated | High-noise environments / AC power. |

- **2-Wire:** The most common. The transmitter is a "parasite" that powers its internal electronics from the residual voltage in the loop. **They are also polarity sensitive. Reversing the leads results in 0 mA and a dead loop.**
- **3/4-Wire:** Use these when the instrument requires more power than 4 mA can provide (e.g., heated sensors, large backlit displays) or when you need total electrical isolation.

## RAW COUNTS AND SCALING

Analog input cards do not "see" milliamps, pressure, or temperature. They see a fluctuating electrical signal that an Analog-to-Digital Converter (ADC) translates into a signed integer—the **Raw Count**.

This integer is the only value the PLC truly "knows." Everything displayed on the HMI after this point is a mathematical construction. Understanding raw counts is the key to isolating faults: if the count is correct, the field hardware is healthy, and the error lies in the software scaling.

**The Digital Multimeter:** > When troubleshooting a suspect loop, your first move—after verifying loop power—is to connect your laptop and view the raw counts directly in the hardware configuration or tag database. This allows you to **Isolate** the fault: if the raw count correctly tracks the loop current, you have proven the field wiring and the input card are healthy. The problem is now officially localized to the scaling software.

Different PLC manufacturers use different raw count ranges for the same 4-20 mA signal. This isn't a problem once you know which platform you're working on, but it matters when you're troubleshooting. If you're expecting Rockwell high-res counts (6,553 to 32,767) and you see Siemens counts (0 to 27,648), you might think the card is broken when it's actually working perfectly.

Always check the card documentation for exact values. The table on Page 102 shows typical ranges, but specific cards may vary.

## Typical Raw Count Ranges by Vendor:

| Platform | 4 mA (0%) | 20 mA (100%) | Notes |
|---|---|---|---|
| Rockwell (Standard) | 3,277 | 16,383 | 1756-IF16 (Unscaled) |
| Rockwell (High-Res) | 6,553 | 32,767 | CompactLogix / 1756-IF8 |
| Siemens S7 | 0 | 27,648 | Normalized Internal Range |
| Schneider / Modicon | 0 | 4,095 | 12-Bit Standard |
| Beckhoff | 6,553 | 32,767 | Terminal-dependent |

*Table Raw Counts and Scaling by Vendor*

⚠ **Critical:** These are approximate. Always check the specific card's documentation for exact values.

## TECH NOTE: BIT RESOLUTION

The "Raw Counts" used by a PLC are determined by the **Bit Resolution** of the Analog-to-Digital Converter (ADC) on the input card. This is essentially how many "slices" the card can chop the 4-20 mA signal into.

- **12-Bit Resolution:** 4,096 slices ($2^{12}$).
- **13-Bit Resolution:** 8,192 slices ($2^{13}$).
- **15-Bit Resolution:** 32,768 slices ($2^{15}$). Standard for many Rockwell cards.

- **Siemens S7 Standard:** Uses a normalized range of 0 to 27,648. This leaves "headroom" above 20 mA so the PLC can still read a signal (up to 22.8 mA) before the value technically "clips" or errors out.

---

**Field Tip:** If you see a raw count of exactly **32,767** or **-32,768** on a Rockwell card, the signal is "railed" (pegged at maximum or minimum). This usually points to a wiring fault or a dead transmitter, not a scaling issue.

---

## Master Scaling Formula

$$EU = \left(\frac{Raw_{In} - Raw_{Min}}{Raw_{Max} - Raw_{Min}}\right) \times (EU_{Max} - EU_{Min}) + EU_{Min}$$

## Field Example: 0–100 PSI (Rockwell High-Res)

$$EU = \left(\frac{19,660 - 6,553}{32,767 - 6,553}\right) \times (100 - 0) + 0 = \mathbf{50 \text{ PSI}}$$

Scaling Formula

Most PLCs use a linear scaling formula:

### Engineering Units = (Raw Counts—Min Counts) ÷ (Max Counts—Min Counts) × (Max EU—Min EU) + Min EU

Example: Pressure transmitter, 0-100 PSI range, Rockwell card

- Raw counts at 4 mA: 6,553
- Raw counts at 20 mA: 32,767
- Min EU: 0 PSI

- Max EU: 100 PSI

If you read 19,660 counts: (19,660—6,553) ÷ (32,767—6,553) × (100—0) + 0 = **50 PSI**

## Common Scaling Errors

| If you see this... | The problem is... | The fix is... |
|---|---|---|
| Counts Good / HMI Wrong | Software Scaling | Match PLC range to Transmitter range. |
| Counts @ 0 or Max / HMI Flat | Hardware/Wiring | Check Loop Power / Field Terminals. |
| 4 mA = 20% on HMI | Config Error | Card set to 0-20mA instead of 4-20mA. |
| Values Swapped (4mA = 100%) | Logic Error | Min/Max values are swapped in PLC. |

💡 **Pro Tip:** When readings are wrong, check raw counts first. If raw counts follow the loop current correctly, the problem is always scaling or configuration—not hardware.

## THE MOST COMMON MISTAKE

New techs see a wrong reading and immediately suspect the transmitter. They pull it, bench-test it, find it's working fine, reinstall it, and the reading is still wrong. Hours wasted.

The raw count check takes 30 seconds and tells you instantly whether the problem is hardware or software. If counts match your meter, the transmitter and wiring are proven good—stop looking at hardware and fix the scaling. If counts don't match, then you troubleshoot the physical signal path.

Raw counts are your decision point. Check them first, every time.

## COMMON FAILURE MODES

| Failure Mode | Symptoms | Field Test |
|---|---|---|
| No Loop Power | 0 mA / PLC Fault | Meter 24VDC at source and transmitter. |
| Open Loop | 0 mA | Check for broken wires or blown fuses. |
| NAMUR Low Fail | ~3.6 - 3.8 mA | Sensor failure; verify process under-range. |
| NAMUR High Fail | ~20.8 - 21 mA | Sensor failure; verify process over-range. |
| Ground Loop | Erratic / Noisy | Ensure shield is grounded at ONE end only. |

## TROUBLESHOOTING WORKFLOW

### Follow the chain

### Power → Transmitter → Wiring → Card → Scaling → Logic

This isn't just a sequence—it's a diagnostic strategy. Each step either confirms or eliminates a link in the chain. If power is good, you've eliminated the supply. If current is correct, you've eliminated power, transmitter, and wiring. If raw counts match current, you've eliminated the card. By the time you reach the HMI and see a wrong value, you know exactly where the fault is.

INSTRUMENTATION & SIGNALS

| Step | Signal Chain | Action |
|---|---|---|
| Symptom | HMI / PLC Scaled Value | Observe the error: Wrong reading, stuck at 0%, pegged at 100%, or fluctuating. |
| Isolate | Loop Current at the Card | Check Loop mA: Use a clamp meter or 250Ω shunt to find actual mA at the PLC card. |
| Test | Loop Current at the Transmitter | Verify Current Match: Does source mA match card mA? Mismatch = wiring fault. |
| Verify | Transmitter Local Display | Compare to Process: Does device display match reality? Disagreement = transmitter fault. |
| Document | Work Order / Calibration Record | Document Truth: Log as-found mA, corrective actions, and verified as-left readings. |

*Table Analog Loop SITVD Chain*

## Step 1: Check Loop Power

Verify 24 VDC at the power supply or card terminals. Meter voltage across the transmitter terminals (typically 12-20 VDC). If **0V**, trace the power path back to the source.

## Step 2: Measure Loop Current

This is the critical physical measurement. Open the loop at a terminal block and insert your meter in series (mA DC), or use a specialized mA clamp-on meter.

## What the current measurement tells you:

| Current (mA) | Diagnostic Meaning | Status / Action |
|---|---|---|
| 0.0 | Open Loop / No Power | **FAULT:** Check wiring & supply |
| <= 3.6 | NAMUR NE43 Fault Signal | **FAULT:** Sensor failure |
| 4.0 | 0% of Process Range | Normal: Lower Range Value |
| 12.0 | 50% of Process Range | Normal: Mid-scale |
| 20.0 | 100% of Process Range | Normal: Upper Range Value |
| > 20.5 | Over-range / Saturated | **WARNING:** Process out of limits |
| Erratic | Signal Noise / Loose Wire | **FAULT:** Check shields & terminals |

## Step 3: Compare Current to PLC Raw Counts

Now you know what the transmitter is sending. Does the PLC agree?

- If current is 12 mA, raw counts should be approximately midpoint (e.g., ~19,660 on a Rockwell card)
- If current and counts agree, the signal path is good
- If counts don't match current, suspect the input card or channel configuration

### Step 4: Compare Raw Counts to Scaled Value

If raw counts match your meter but the HMI is wrong, the fault is isolated to the **Scaling Software**. Verify the input range (4-20mA vs 0-20mA), check scaling parameters ($Min/Max$), and ensure the HMI isn't applying a secondary multiplier or offset.

### Step 5: Verify Against Field Reference

Finally, prove the loop against a physical standard.

If the reference (e.g., test gauge, ice bath, bucket test, or sight glass), transmitter output, PLC raw counts, and HMI all agree, the loop is healthy. If any link doesn't match, you've localized the fault.If the reference, transmitter output, PLC raw counts, and HMI display all agree, the loop is healthy.

## USING A LOOP CALIBRATOR

## HART PROTOCOL

Most loop-powered transmitters installed in the last 25 years are HART-capable. HART (Highway Addressable Remote Transducer) is a digital communication protocol that rides on top of the standard 4-20 mA loop signal without disturbing it. The transmitter simultaneously outputs its analog current and transmits digital data—device status, process variables, diagnostic information, and configuration parameters—on the same two wires.

Why this matters for troubleshooting: a HART communicator (handheld or PC-based) lets you talk directly to the transmitter without breaking the loop. You can read the transmitter's self-reported process value, its output current, its diagnostic status, its range and span configuration, and any fault codes—all while the loop remains live and the process continues running.

Three field uses you need to know:

1. Verify transmitter configuration without disconnection. Range, engineering units, damping settings, and sensor type are all readable via HART. If the transmitter is ranged 0-500 PSI but the PLC scaling is set for 0-100 PSI, HART shows you the mismatch in seconds.
2. Read transmitter diagnostics. A HART-capable transmitter reports its own health. Sensor degradation, loop current out of range, temperature extremes, and electronics faults are flagged in the device status register—information that never appears on the HMI.
3. Trim zero and span in the field. Rather than disconnecting the transmitter and taking it to the shop, zero and span trimming can be done live via HART with a known reference applied.

How to identify a HART transmitter: look for the HART logo on the nameplate, or check the model number against the manufacturer's documentation. If the transmitter is from any major process instrumentation vendor (Emerson, Endress+Hauser, ABB, Yokogawa, Honeywell) and was manufactured after 2000, assume HART unless proven otherwise.

Pro Tip: You don't need a dedicated HART communicator to get started. Many calibrators include HART capability, and PC-based HART interfaces are available for under $300. If your plant has HART transmitters and no communicator on the shelf, that's a gap worth closing before your next difficult loop diagnosis.

A loop calibrator is your second-most important tool after your multimeter.

It allows you to source current to test the signal path, simulate a transmitter to prove the wiring and card are healthy, and provide 24V loop power for bench-testing instruments.

## The Substitution Test

When you suspect a bad transmitter:

1. Disconnect the transmitter from the loop.
2. Connect your calibrator in place of the transmitter.
3. Source 12.0 mA (50%).
4. The Result: If the PLC shows 50%, the signal path (wiring/card) is healthy and the transmitter has failed. If it shows 0%, the fault is in the path.

# SHIELD GROUNDING

### ANALOG OUTPUTS (AO)

This chapter has focused on analog inputs—the PLC reading the field. Analog output cards work the opposite direction: the PLC writes a value, the card converts it to a 4-20 mA signal, and that signal drives a control valve positioner, a VFD speed reference, or a proportional valve.

Troubleshooting an analog output follows the same chain in reverse:

PLC Output Value → Card → Wiring → Actuator

Step 1: Confirm the PLC is writing the intended value. Go online and read the output tag. If the tag value is wrong, the problem is in the control logic, not the hardware.

Step 2: Measure the card's actual output current at the terminal. If the PLC tag says 50% (12 mA) but the terminal measures 4 mA or 0 mA, the card or channel has failed.

Step 3: Measure current at the actuator terminals. If the card is outputting 12 mA but the actuator terminal shows 0 mA, there is a break in the wiring between them.

Step 4: Verify actuator response. A control valve positioner that receives correct current but doesn't move has a mechanical or positioner problem, not a signal problem.

Common analog output failure: a card channel that outputs a fixed value regardless of the PLC tag—usually the result of a failed D/A converter on that channel. The other channels on the same card continue working normally.

## VOLTAGE SIGNALS: 0-10V AND VARIANTS

Not every analog signal is a current loop. HVAC equipment, some OEM machinery, and servo/VFD speed references frequently use voltage signals: 0-10 VDC, 0-5 VDC, or ±10 VDC. The troubleshooting approach is the same—verify at the source, trace the wiring, confirm at the destination—but the measurement technique differs. Voltage signals are high-impedance; you measure across the signal terminals with the circuit live, not in series. You do not break the loop to measure a voltage signal. The failure modes also differ: a voltage signal output can be "loaded down" by a low-impedance destination (wrong input card type), causing a reading lower than expected even though the source is outputting the correct voltage.

If you encounter a signal you can't characterize, check the transmitter nameplate and the input card documentation before assuming 4-20 mA. Connecting a 0-10 VDC output to a 4-20 mA input card will produce incorrect readings and may damage the card.

Analog signals are low-level (milliamps). They're susceptible to picking up noise from nearby power cables, VFDs, and other electrical equipment. Shielded cable helps, but only if grounded correctly.

**The Rule: Ground the shield at ONE end only—typically the PLC end.**

If you ground the shield at both ends, you create a ground loop. Current flows through the shield (because the grounds at each end are at slightly different potentials), and this current creates noise that couples into your signal.

**Proper shield termination:**

- At the PLC end: Connect shield to the instrument ground terminal
- At the transmitter end: Cut the shield short and tape it back. Do not connect it to anything.

**Exception:** Some installations require both ends grounded for safety reasons. In that case, use a shield grounding kit that includes a capacitor to block DC ground loop current while still providing high-frequency shielding.

## QUICK FIELD VERIFICATION

*No calibrator? Use a physical reference.*

- **Pressure:** Install a tee-gauge and compare to transmitter output.
- **Temperature:** Use a thermocouple simulator, ice bath ((0°C / 32°F), or boiling water (100°C / 212°F).
- **Flow:** Perform a timed bucket test (e.g., 5 gallons in 30 sec = 10 GPM).
- **Level:** Use a dipstick, tape measure, or sight glass.

*If the reference and the transmitter agree, and the PLC doesn't, the problem is downstream (wiring, card, or scaling).*

## DECISION TREE REFERENCE

See **Appendix C: Analog Loop (4-20 mA) Troubleshooting** for a step-by-step flowchart you can photocopy and laminate.

💡 **Pro Tip:** Raw counts never lie. If counts track your measured current, the transmitter and wiring are fine—fix the scaling.

# CHAPTER 14
# TEMPERATURE SENSORS

## TEMPERATURE SENSORS

Temperature is one of the most common measurements in industrial processes.

Ovens, furnaces, heat exchangers, refrigeration systems, chemical reactors, extruders, molding machines, environmental chambers—they all depend on accurate temperature data to operate correctly and safely.

But temperature sensors fail in subtle ways. Unlike a pressure transmitter that either works or doesn't, a temperature sensor can give you readings that are *believable but wrong*. The display shows a number that looks reasonable, so nobody questions it—until the product is ruined, the batch is rejected, or the equipment is damaged.

This chapter covers the three main types of temperature sensors you'll encounter in the field: RTDs, thermocouples, and infrared sensors—how they work, how they fail, and how to verify they're telling the truth.

## SENSOR TYPES AT A GLANCE

The industrial world uses many temperature sensor technologies—thermistors, bimetallic elements, filled-system sensors, fiber optic, and more.

In practice, RTDs, thermocouples, and infrared sensors account for the overwhelming majority of what you'll encounter on the plant floor.

*Common 3-wire RTD*

. . .

Thermistors—semiconductor-based temperature sensors common in HVAC equipment, refrigeration controls, and OEM machinery—are not covered in this chapter. They operate on a different principle (non-linear resistance change, typically NTC) and require different input cards and measurement techniques. Thermistor troubleshooting reference material is available at BootsOnTheGroundTech.com. That's what this chapter covers.

## RTDs (RESISTANCE TEMPERATURE DETECTORS)

An RTD uses the principle that the electrical resistance of a metal changes predictably with temperature. The most common type is the Pt100—a platinum element with a resistance of exactly 100 ohms at 0°C (32°F). As temperature increases, resistance increases. The relationship is nearly linear and extremely repeatable—which is why RTDs are the go-to sensor when accuracy and stability matter.

The table below shows the most common RTD types you'll encounter in industrial settings. Pt100 and Pt1000 dominate—Pt100 for legacy systems and high-temperature applications, Pt1000 for newer installations where higher resistance provides better noise immunity over long cable runs.

The resistance values listed are at 0°C (32°F). As temperature increases, resistance increases proportionally. A Pt100 at 100°C reads 138.5 Ω. A Pt1000 at the same temperature reads 1,385 Ω—exactly 10 times higher.

When you're troubleshooting an RTD, knowing the expected resistance at a given temperature lets you verify the sensor is working before you blame the transmitter, wiring, or card.

The table below shows resistance values at 0°C for the most common industrial RTDs. When troubleshooting, knowing the expected resistance at a given temperature lets you verify the sensor before blaming the transmitter or card.

## Common RTD Types:

| RTD | Material | Ω @ 0°C | Range | Use | Notes |
|---|---|---|---|---|---|
| Pt100 | Platinum | 100 Ω | -200°C to +850°C | General industrial HVAC Food/Pharma | Most common. Highly accurate; IEC 60751 std. |
| Pt1000 | Platinum | 1000 Ω | -200°C to +850°C | Low-current; Long cable runs | 10X resistance reduces lead-wire error; common in Europe. |
| Ni100 | Nickel | 100 Ω | -60°C to +180°C | HVAC, Building Automation | More sensitive in narrow ranges; Not interchangeable with Pt100. |
| Ni120 | Nickel | 120 Ω | -60°C to +180°C | North American HVAC (Honeywell/JCI) | Common in US building controls; same limitations as Ni100. |
| Cu10 | Copper | 10 Ω | -200° to +260°C | Motor windings, Transformers | Very linear; low resistance requires exact lead compensation. |

*Table Common RTD Types*

## RTD WIRING CONFIGURATIONS

RTDs come in 2-wire, 3-wire, and 4-wire configurations. If you land a 3-wire RTD as a 2-wire (leaving one wire disconnected), you'll get readings that are consistently too high. If you land a 2-wire RTD on a 3-wire input (jumper two terminals together), the card may compensate incorrectly. The difference matters because lead wire resistance adds measurement error—and how the input card compensates depends entirely on which configuration it's set for.

⚠ **Shop-Floor Wisdom: Shop-Floor Wisdom:** If you land a 3-wire RTD as a 2-wire (leaving one wire disconnected), you'll get readings that are consistently too high. If you land a 2-wire RTD on a 3-wire input (jumper two terminals together), the card may compensate incorrectly. Check the wiring before blaming the sensor.

## THERMOCOUPLES

A thermocouple uses the Seebeck effect: when two dissimilar metals are joined at a junction, a small voltage is generated that varies with temperature. This voltage is tiny—measured in millivolts—but it's predictable and repeatable.

The hot junction is the sensing point. The cold junction (reference junction) is where the thermocouple wires connect to the measurement device.

# Temperature Sensors

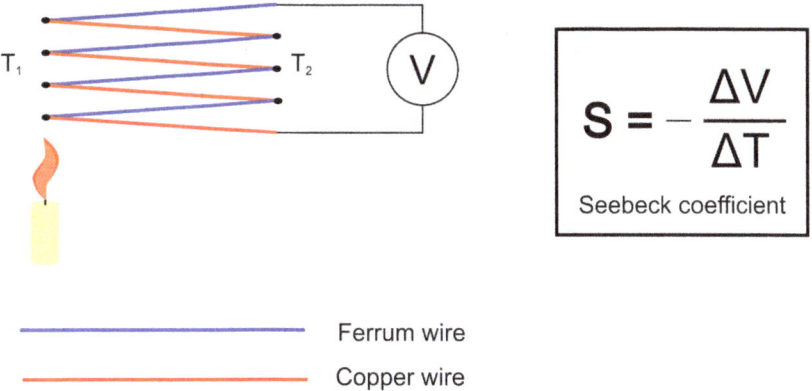

The device measures the voltage difference and calculates temperature based on known characteristics of the metal combination.

## Common Thermocouple Types:

Type K is the workhorse. If you're not sure what type a thermocouple is and you're in a general industrial environment, guess Type K. You'll be right most of the time.

| Type | Metals | Range (°C) | Output (mV) | Color (US) | Notes |
|---|---|---|---|---|---|
| K | Chromel/Alumel | -200 – +1250 | 41.2 | YELLOW | Most common general purpose |
| J | Iron/Constantan | -40 – +750 | 52.7 | BLACK | Reducing atmospheres |
| T | Copper/Constantan | -200 – +350 | 20.9 | BLUE | Cryogenic / Food safety |
| E | Chromel/Constantan | -200 – +900 | 76.4 | PURPLE | Highest output voltage |
| N | Nicrosil/Nisil | -270 – +1300 | 47.5 | ORANGE | Stable K alternative |
| R | Plat/Rhodium | 0 – +1450 | 18.7 | WHITE | High Temperature / Expensive |
| S | Plat/Rhodium | 0 – +1450 | 18.7 | WHITE | High Temperature / Expensive |

CRITICAL: EXTENSION WIRE MUST MATCH

Thermocouple extension wire must be made of the same metals as the thermocouple (or thermoelectrically equivalent alloys). If you splice copper wire into a thermocouple circuit, you create additional junctions that corrupt the measurement. Look for mismatched colors at junction boxes—that's the tell.

## INFRARED (NON-CONTACT) SENSORS

Infrared (IR) sensors measure thermal radiation emitted by a surface. This non-contact approach is ideal for moving machinery, high-voltage equipment, or hazardous environments where physical probes are impractical or dangerous.

### Understanding the D:S Ratio

The Distance-to-Spot (D:S) ratio is the most critical physical constraint of an IR sensor. It defines the diameter of the measurement spot based on the distance from the target.

- Averaging Error: An IR sensor measures the average temperature of everything in its field of view.
- 12:1 Ratio: At a 12-inch distance, you are measuring a 1-inch spot. At 12 feet, you are measuring a 12-inch circle.
- Background Interference: If your target is a 2-inch pipe but your spot is 12 inches, the sensor will average the pipe with the background wall or oven.

To get an accurate reading, the target must be larger than the sensor's spot size. If you can't get closer, you need a sensor with a higher D:S ratio (e.g., 30:1 or 50:1) to narrow the beam.

*Typical pyrometer*

## The Emissivity Factor

The critical variable is *Emissivity* $\varepsilon$—a measure of how efficiently a surface radiates energy. It ranges from *0* to *1*. If the sensor setting does not match the target surface, the reading will be "believable but wrong."

| Surface Type | Emissivity Range | Example Materials |
|---|---|---|
| High Emissivity | 0.90 to 0.98 | Matte black paint, asphalt, brick, wood, cloth, rubber. |
| Medium Emissivity | 0.40 to 0.80 | Oxidized metals, plastics, paper, food products, water. |
| Low Emissivity | 0.05 to 0.90 | Polished stainless steel, aluminum foil, chrome, gold. |

**Example:** A polished stainless steel tank at 100°C might read 50°C if the sensor is set for emissivity = 0.95 (matte black). Correct the emissivity setting (~0.15 for polished stainless) or apply matte black tape to the measurement spot.

### Why Emissivity Settings Matter

Most IR sensors ship with a default emissivity of 0.95, which works for many industrial surfaces. But if you're measuring polished stainless steel ($\varepsilon$ = 0.15) with that setting, your reading will be drastically wrong—often 50-100°F low.

The fix is simple: verify the target surface material, look up the correct emissivity value, and program it into the sensor. Wrong emissivity is the #1 cause of "believable but wrong" IR temperature readings.

When you commission a new IR sensor, always verify the emissivity setting matches the target surface. Don't assume the factory default is correct. Five minutes of setup prevents months of chasing phantom temperature problems that aren't real.

# Temperature Sensors

| Failure Mode | Symptom | Cause | Fix |
|---|---|---|---|
| Dirty Lens | Reading drifts low or stays frozen. | Dust, oil, or steam blocking the optical path. | Clean lens with air or soft cloth; use air purge. |
| ε Mismatch | Believable but wrong (e.g., reads 40°C at 100°C). | Emissivity setting does not match target material. | Adjust ε setting; apply matte tape to target. |
| Background Noise | Erratic spikes or "hot" offsets. | Hot objects (ovens/lights) reflected in shiny target. | Shield the sensor; change angle to avoid reflections. |
| Spot Size Error | Average of target + background. | Sensor is too far; "sees" more than the target area. | Move sensor closer; check Distance-to-Spot (D:S) ratio. |

*Table Common Emissivity Faults and Fixes*

## THE "BELIEVABLE BUT WRONG" PROBLEM

This is the most dangerous failure mode for temperature sensors. The reading is in the expected range, it responds to changes, nothing looks obviously broken—but it's wrong. The process runs confidently on bad data.

### The Mechanics of Deception

A "believable but wrong" reading occurs because the circuit remains closed. The PLC sees a valid signal and scales it according to programmed parameters, unaware the input is corrupted.

### Common Scenarios

Thermocouple type mismatch is the classic example. A Type K thermocouple installed in a Type J input reads 10-50°F wrong—close enough to seem plausible. RTD wiring configured as 2-wire instead of 3-wire reads consistently high due to lead resistance. Emissivity set to 0.95 on polished metal reads 50-100°F low.

In every case, the signal path is intact, the sensor functions, and the PLC displays a number. But the number is fiction.

| Cause | What Happens | How to Catch It |
|---|---|---|
| Wrong TC type configured | Reading offset 10–50°C | Check wire color, label, and input config — must all match |
| Reversed TC polarity | Reading moves wrong direction or is offset | Warm the sensor — reading must go UP |
| Wrong extension wire | Corrupted measurement at every splice point | Trace wiring; any non-matched wire is suspect |
| RTD wiring error | Reads high (3-wire as 2-wire) or erratic | Meter resistance; compare to expected value for ambient temp |
| Sensor not bottomed in thermowell | Reads between process and ambient | Push sensor down; see if reading changes |
| Degraded/drifted sensor | Stable, but consistently offset | Compare to ice bath, boiling water, or calibrated reference |

*Table Believable but Wrong Problems*

## COMMON FAILURE MODES

| Failure | Sensor Type | Symptom | Cause | Field Test |
|---|---|---|---|---|
| Open circuit | RTD | Infinite resistance / off-scale high or fault | Fatigue, corrosion, damage | Meter resistance at terminals |
| Open circuit | TC | Reads room temperature (cold junction) or fault | Fatigue, corrosion, damage | Meter continuity |
| Short circuit | RTD | Near-zero resistance / off-scale low or fault | Damaged, water ingress, crushed cable | Meter resistance; compare to expected for ambient |
| Short circuit | TC | Reads cold junction temp regardless of process | Damaged, water ingress, crushed cable | Meter continuity; leads should not be shorted |
| High resistance | RTD | Reads too high (extra resistance = extra temp) | Loose terminals, corrosion, poor crimps | Wiggle test — noisy/jumping values = bad connection |
| High resistance | TC | Erratic or noisy readings | Loose terminals, corroded connections, poor crimps | Wiggle test — noisy/jumping values = bad connection |
| Noise pickup | TC | Readings fluctuate; noise correlates with VFDs / welders | Unshielded cable, cable routed near power wiring | Check trend — bouncy signal on stable process = noise |
| Sensor drift | Both | Consistently offset from reality — stable but wrong | High-temp exposure (TC), thermal cycling (RTD) | Compare to known reference (ice bath / Boiling water) |

*Table Common Temperature Failure Modes*

# THE SITVD TEMPERATURE WORKFLOW

| Step | Description | Temperature Application | Actionable Field Check |
|---|---|---|---|
| Symptom | Identify the failure mode. | Is the reading erratic, frozen, offset, or pegging high/low? | Trend the value; compare process behavior to displayed reading. |
| Isolate | Break the loop to determine fault. | Probe vs. Lead-Wire/Card error. | Disconnect sensor leads; use a simulator or verify card counts directly. |
| Test | Check sensor's physical health. | Meter resistance (RTD) or continuity (TC). | Use the Pt100 Resistance Reference Table or verify mV changes when heated. |
| Verify | Check installation and logic. | The "Triple Match": Sensor type, input type, and scaling parameters. | Audit the physical installation (Thermowell bottomed? Paste applied? Damage?). |
| Document | Close out scaling/raw data lies. | Ensure the signal translation is correct. | Verify Raw Counts match the loop measurement and check for HMI offsets. |

*Table SITVD Workflow*

## S—Symptom

- Identify the failure mode.
- Is the reading erratic, frozen, offset, or pegging high/low?
- Trend the value; compare process behavior to displayed reading.

**Example:** A temperature reading that stays at 25°C while the heater is clearly red-hot is a "Frozen" symptom, suggesting a configuration mismatch.

## I—Isolate

- Break the loop to determine fault.
- Probe vs. Lead-Wire/Card error.
- Disconnect sensor leads; use a simulator or verify card counts directly.

**Example:** Disconnect the RTD at the head. If the PLC still shows 150°C, the "lie" is in the wiring or the input card, not the sensor.

## T—Test

- Check sensor's physical health.
- Meter resistance (RTD) or continuity (TC).
- Use the Pt100 Resistance Reference Table or verify mV changes when heated.

**Example:** A Pt100 RTD at room temperature should meter ~109.7$\Omega$. If you see 0$\Omega$ or Infinite, the element is physically destroyed.

## V—Verify

- Check installation and logic.
- The "Triple Match": Sensor type, Input type, and scaling parameters.
- Audit the physical installation (Thermowell bottomed? Paste applied? Damage?).

**Example:** Ensure a Type K thermocouple is actually connected to a Type K input card. A mismatch here is the #1 cause of "Believable but Wrong" data.

## D—Document

- Close out scaling/raw data lies.
- Ensure the signal translation is correct.
- Verify Raw Counts match the loop measurement and check for HMI offsets.

**Example:** If the PLC sees 138.5$\Omega$ (correct for 100°C) but the HMI shows 105°C, document the scaling error and update the HMI tag parameters.

## THERMOWELLS

Most temperature sensors in process environments are installed in thermowells—protective metal tubes that isolate the sensor from the process fluid while allowing heat transfer.

| Problem | Effect | How to Check |
|---|---|---|
| Coating / Fouling | Insulates well from process; slow response and offsets. | Remove sensor; inspect well exterior for scale or buildup. |
| Sensor not bottomed | Reads between process and ambient (air gap error). | Push sensor to bottom; watch for an immediate reading change. |
| Wrong length | Doesn't measure representative process temperature. | Verify insertion spec against pipe size and flow profile. |
| Internal debris | Sensor cannot seat properly; poor heat transfer. | Inspect bore with a light; clean or replace the well. |

*Table Thermowell Common Issues*

## TEMPERATURE TRANSMITTERS (HEAD-MOUNTED)

Many industrial temperature installations do not connect the RTD or thermocouple directly to a PLC input card. Instead, a small transmitter—DIN-rail mounted or fitted directly into the sensor connection head—converts the sensor signal to a standard 4-20 mA loop output. From the PLC's perspective, it looks like any other analog loop: loop power, milliamp signal, raw counts, scaling.

This adds a failure node to the SITVD chain that the standard workflow doesn't account for. If you disconnect the sensor and measure the transmitter's 4-20 mA output, you are measuring the transmitter—not the sensor. A transmitter with a failed amplifier can output a fixed or saturated milliamp value even with a good sensor attached.

**Field identification:** look for a small housing in the sensor terminal head (aluminum or plastic box inside the connection head), or a DIN-rail mounted device labeled as a temperature transmitter or "T/T." The transmitter will have its own sensor-type configuration—Type K, Type J, Pt100, etc.—that must match the sensor, independent of any PLC input card configuration.

**SITVD adjustment:** when a head-mounted transmitter is in the circuit, add a step between I (Isolate) and T (Test): disconnect the transmitter output and verify the 4-20 mA signal directly. If the output is wrong with a known-good sensor connected, the transmitter has failed.

## INPUT DEVICE CONSIDERATIONS

Temperature input cards handle a lot behind the scenes—and misconfiguration at this layer causes readings that are wrong despite a perfectly healthy sensor.

| Function | What It Does | What Goes Wrong |
|---|---|---|
| Cold Junction Comp (CJC) | Measures terminal temp and offsets TC reading. | Bad CJC sensor → TC readings consistently offset. |
| Linearization | Applies correction curve for TC non-linearity. | Wrong TC type selected → wrong curve → wrong reading. |
| Filtering / Averaging | Smooths noise over time. | Too aggressive → reading lags process by several seconds. |
| Open Circuit Detection | Flags a fault when sensor wire breaks. | Not all cards do this — know your card's behavior. |

## REAL-WORLD EXAMPLE

**Problem:** Extruder Zone 3 running 40°C hotter than setpoint. Heater bands checked out fine. PID loop output normal.

**Symptom:** HMI shows 240°C on Zone 3. Setpoint is 200°C. Operator increased cooling dwell time to compensate. Product quality still inconsistent.

**Troubleshooting:**

1. Check configuration: Zone 3 input card configured for Type J. Physical sensor label reads "Type K." Previous maintenance tech replaced the probe but didn't update the input configuration.
2. Cross-check: At 200°C, Type J and Type K outputs differ by approximately 40°C—exactly the observed offset.
3. Fix: Updated input configuration to Type K. Zone 3 temperature immediately dropped to 200°C and tracked setpoint normally.

**Total time:** 18 minutes.

**Root cause:** Sensor type mismatch—probe replaced, configuration not updated. The reading was believable, stable, and wrong.

## ⚡ PRO TIP: THE QUICK-CHECK

If you suspect a mismatch, pull the sensor and put it in a boiling water bath (100°C).

- A Type J configured as Type K will read ~120°C.
- A Type K configured as Type J will read $ 80° C.

———

## QUICK FIELD VERIFICATION TABLE

| Test | Method | Expected Result |
|---|---|---|
| Ice Bath | Crushed ice + water, sensor immersed | 0°C / 32°F |
| Boiling Water | Rolling boil at sea level | 100°C / 212°F |
| Room Temp | Sensor in ambient air | ~20-25°C / 68-77°F |
| Body Heat | Closed fist around sensor | ~35-37°C / 95-99°F |
| RTD Resistance | Ohmmeter at input terminals | 100.0 Ω @ 0°C / ~110 Ω @ 25°C |
| TC Continuity | Ohmmeter through junction | Near 0 Ω |

## DECISION TREE REFERENCE

A temperature sensor troubleshooting decision tree is available at BootsOnTheGroundTech.com for download and printing.

💡 **Pro Tip:** When a temperature reading is "believable but wrong," check sensor type, wiring configuration, and input settings before assuming the sensor has failed. The hardware is usually fine—the configuration is wrong.

⚡ **Shop-Floor Wisdom:** If a thermocouple reads "believable but wrong," check polarity and type before replacing. A Type J in a Type K input will give you a number—just not the right one.

# CHAPTER 15
# PRESSURE SENSORS

## PRESSURE SENSORS

Pressure is everywhere in industrial systems. Hydraulic circuits, pneumatic actuators, coolant loops, steam lines, refrigeration systems, chemical reactors, fuel supplies, lubrication systems—all depend on pressure being where it should be.

A pressure sensor converts mechanical force into an electrical signal. That signal tells the PLC whether pressure is in range, whether a pump is running, whether a filter is clogged, or whether a vessel is about to rupture. When pressure sensors fail or read wrong, the consequences range from nuisance alarms to catastrophic equipment damage.

This chapter covers the main types of pressure sensors, how they work, how they fail, and how to verify their readings in the field.

## THE PHYSICAL REALITY OF PRESSURE

In the field, pressure is more than just a number on an HMI—it is stored energy. Every PSI represents potential force acting against seals, gaskets, and vessel walls. A sensor that drifts high might trigger a nuisance relief valve pop; a sensor that drifts low could allow a system to over-pressurize until a mechanical failure becomes the "safety" of last resort.

Troubleshooting these devices requires a respect for the medium. Whether you are dealing with a standard 4-20mA transmitter or a simple pressure switch, the goal remains the same: verify that the electrical representation in the PLC matches the physical reality in the pipe.

## PRESSURE MEASUREMENT BASICS

Before diving into sensor types, understand what you're measuring. The same process can be described three different ways depending on the sensor type installed—and getting this wrong produces readings that are believable but completely incorrect.

**Gauge pressure (PSIG):** Measured relative to atmospheric pressure. A tire inflated to 30 PSIG is 30 PSI above the surrounding atmosphere. Most industrial sensors measure gauge pressure.

**Absolute pressure (PSIA):** Measured relative to a perfect vacuum. Atmospheric pressure at sea level is approximately 14.7 PSIA. Used in vacuum systems and altitude-sensitive applications. A gauge sensor in a vacuum system will peg at zero even if significant vacuum is present.

**Differential pressure (PSID):** The difference between two pressure points. Used for flow measurement (orifice plates), filter monitoring, and level measurement in pressurized vessels.

### Pressure Units

| Type | How It Works | Strengths | Limitations | Applications |
| --- | --- | --- | --- | --- |
| Strain-Gauge | Diaphragm flexes; gauges change in resistance. | Rugged, wide range, stable, accurate. | Susceptible to overpressure/fatigue. | General industrial. |
| Piezoelectric | Pressure on crystal generates charge. | Extremely fast; dynamic/pulse sensing. | Cannot measure static pressure. | Engine testing, blast monitoring. |
| Capacitive | Diaphragm is one plate of a capacitor. | Very sensitive; stable; low vibration drift. | Temperature/humidity sensitive. | Low-pressure HVAC, clean rooms. |
| MEMS | Etched silicon diaphragm on a chip. | Small, fast, low cost at volume. | Less robust; sensitive to EMI. | Automotive, medical, OEM kits. |

Field Note: In heavy industrial environments, you're almost always dealing with a strain-gauge transmitter. Piezoelectric sensors are special-purpose—if someone installed one expecting a steady reading, that's a specification error, not a sensor failure. (See "The Piezoelectric Exception" below.)

## THE PIEZOELECTRIC EXCEPTION

This one trips people up enough to deserve its own callout. Piezo sensors respond to *changes* in pressure—not static pressure. The charge naturally dissipates over time, so the output decays back to zero even if pressure remains constant. If you're troubleshooting a sensor that "works for a moment then drifts back to zero," that's normal piezo behavior. They show the spike, then decay.

If someone installed a piezo sensor expecting it to hold a steady reading, the sensor isn't broken—it's the wrong tool.

## SNUBBERS AND DIAPHRAGM SEALS

Snubbers protect transmitters in pulsating applications—reciprocating pumps, hydraulic systems, anywhere pressure spikes are common. A sintered metal disc or orifice restricts flow, smoothing out pulses before they reach the sensor. Trade-off: more dampening equals slower response. Find the balance between protecting the transmitter and maintaining acceptable response time.

Diaphragm seals isolate the transmitter from corrosive, viscous, or high-temperature process fluids. A flexible diaphragm separates the process fluid from the sensor, with fill fluid (typically silicone oil) transmitting pressure across. They add cost, and fill fluid expands and contracts with temperature—but they're the right answer when the process would destroy a direct-mount transmitter.

## COMMON FAILURE MODES

| Failure | Symptom | Cause | Prevention / Fix |
|---|---|---|---|
| Overpressure | Permanent zero shift; nonlinear response. | Water hammer, pump deadhead, pressure spikes. | Install snubber; verify sensor range vs. max system spikes. |
| Corrosion | Erratic readings; leakage from body. | Fluid incompatible with wetted materials. | Use diaphragm seals for corrosive or viscous fluids. |
| Zero Drift | Doesn't return to 4 mA at zero pressure. | Temperature cycling, age, prior overpressure. | Recalibrate; if drift recurs, replace the sensor. |
| Span Drift | Reading compressed or expanded vs. actual. | Environmental stress; aging electronics. | Recalibrate against a known pressure reference. |
| Electrical | No output (0 mA) or fixed/erratic output. | Failed signal conditioning electronics. | Apply pressure; if output is dead but mechanical is OK, replace. |
| Temp Effects | Shifts that correct when temp stabilizes. | Thermal expansion beyond compensated range. | Use rated environmental housings; shade from radiant heat. |

*Table Common Pressure Sensor Failures*

## TROUBLESHOOTING WORKFLOW

Follow the chain: Power → Transmitter → Impulse Line → Reference → Scaling

### Step 1: Verify Power

Most pressure transmitters are 2-wire, loop-powered devices. Meter voltage across the transmitter terminals—should see 10-20 VDC at the transmitter (some voltage drops in the loop). Zero volts means trace the power path: fuse, wiring, power supply.

### Step 2: Measure Loop Current

With power confirmed, measure mA in series (or use a clamp-on mA meter).

| mA Reading | What It Means |
|---|---|
| 0 mA | Open circuit — wiring break or transmitter failure. |
| 3.6–3.9 mA | Transmitter internal fault / alarm condition. |
| 4.0 mA | Zero pressure (or calibrated zero). |
| 4–20 mA tracking | Transmitter functioning — compare value to actual pressure. |
| 20+ mA | Overrange condition or transmitter electronics fault. |
| Fixed value | Electronics failed or impulse line is physically blocked. |

*Table Loop Current vs Current*

## Step 3: Compare to a Reference

Connect a calibrated test gauge to the same pressure tap (or tee into the impulse line). This is the definitive test. If the test gauge agrees with the transmitter, the transmitter is correct and the problem is downstream (wiring, scaling, HMI). If they disagree, the problem is the transmitter or the impulse line.

## Step 4: Check the Impulse Line

If the test gauge at the transmitter shows different pressure than the process, the impulse line is plugged, leaking, or has trapped air or condensate. Crack the fitting carefully, bleed, and re-test.

## Step 5: Check Raw Counts and Scaling

If transmitter output is correct but the displayed value is wrong, compare loop current to PLC raw counts, verify scaling parameters (0-100 PSI? 0-500 PSI?), and check for HMI-level scaling or offsets.

## FIELD VERIFICATION METHODS

| Method | When to Use | Procedure | Notes |
|---|---|---|---|
| Test Gauge (Tee) | First field check. | Tee a calibrated gauge into the line; compare to transmitter. | If they agree, the transmitter is healthy. |
| Hand Pump | Bench test or isolated loop. | Isolate/vent transmitter; apply 0%, 50%, 100% of range. | Confirms transmitter independently of process. |
| Loop Calibrator | Prove wiring / PLC. | Disconnect transmitter; inject 4/12/20 mA to the PLC. | Proves the signal path is not the "liar." |
| Deadweight Tester | High-accuracy lab work. | Apply precisely known pressure using calibrated weights. | Laboratory tool; rarely used for field triage. |

## REAL-WORLD EXAMPLES

### Example 1: The Sluggish Reading

**Problem**: Pressure reading on a hydraulic system takes 30+ seconds to respond to changes.

**Finding**: Transmitter connected via 10 feet of 1/8" impulse tubing with a partial plug from hydraulic fluid contamination.

**Fix**: Removed and cleaned the impulse line. Installed a snubber rated for the application to dampen pulsation without excessive restriction.

**Time**: 25 minutes.

### Example 2: The Disappeared Pressure

**Problem**: Coolant pressure reads zero but the pump is running and coolant is flowing.

**Investigation**: Transmitter powered, outputting exactly 4.0 mA. Test gauge at the transmitter showed 0 PSI. Test gauge at the process connection showed 45 PSI.

**Finding**: Impulse line completely plugged with scale. Pressure at the process never reached the transmitter.

**Fix**: Replaced impulse tubing. Added a blowdown valve for periodic cleaning.

**Time**: 30 minutes.

### Example 3: The Reading That Made No Sense

**Problem**: Reactor pressure showing 250 PSI at startup before anything was running.

**Investigation**: Transmitter was 0-500 PSI range. Loop current was 12 mA (50% = 250 PSI). Applied 0 PSI with hand pump—output still 12 mA.

**Finding**: Transmitter had been over-pressured during a previous event. Diaphragm was permanently deformed, creating a 250 PSI zero offset.

**Fix**: Replaced transmitter. Investigated cause of overpressure event and installed a pressure relief device.

## QUICK FIELD VERIFICATION

| Signal % | Expected mA | Raw Count (Rockwell) | Meaning |
|---|---|---|---|
| 0% | 4.0 mA | $\cong$ 6,553 | Calibrated zero |
| 25% | 8.0 mA | $\cong$ 13,107 | 1/4 Scale |
| 50% | 12.0 mA | $\cong$ 19,661 | Midpoint check |
| 75% | 16.0 mA | $\cong$ 26,215 | 3/4 Scale |
| 100% | 20.0 mA | $\cong$ 32,767 | Full scale |

💡 **Pro Tip:** When troubleshooting pressure, a simple mechanical test gauge is your best friend. If the gauge reads correctly but the transmitter doesn't, you've isolated the problem to the transmitter or its impulse line.

⚡ **Shop-Floor Wisdom:** If a pressure reading is sluggish or stuck, check the impulse line before condemning the transmitter. A plugged line is a $10 fix. A new transmitter is $500.

Pressure measurement is fundamental across all industrial automation—whether you're monitoring hydraulic systems on a press line, compressed air on a packaging machine, or vacuum on a pick-and-place operation. When a pressure reading is wrong, follow the signal chain systematically: verify with a test gauge, check the impulse line, meter loop current, compare raw counts, verify scaling. Most pressure problems are impulse line issues or configuration errors—actual transmitter failures are rare.

# CHAPTER 16
# FLOW METERS

## FLOW METERS

Flow measurement tells you how much fluid is moving through a pipe—gallons per minute, liters per hour, cubic feet per day. It's essential for batching, blending, billing, cooling systems, chemical dosing, and process control.

Flow meters come in many varieties, each suited to different fluids and applications. As a troubleshooter, you don't need to be an expert in flow meter design. You need to know what type you're dealing with, how it fails, and how to verify whether the reading is believable.

This chapter covers the most common flow meter types you'll encounter in discrete and process manufacturing.

## FLOW MEASUREMENT BASICS

Volumetric vs. Mass Flow: Volumetric flow measures volume per unit time (GPM, liters/min, cubic feet/hour)—what most flow meters measure. Mass flow measures mass per unit time (lb/min, kg/hour), which matters when density changes with temperature or pressure, or when billing by weight. For most water and stable liquid applications, volumetric flow is fine. For gases, steam, or fluids with variable density, mass flow may be required.

Flow Rate vs. Totalized Flow: Flow rate is the instantaneous measurement (GPM right now). Totalized flow is the accumulated measurement (total gallons over time). Most meters provide both. If the totalizer is wrong but the rate looks correct, suspect the totalizer configuration—pulses per gallon or scaling factor.

## METER TYPES AT A GLANCE

| Type | Principle | Moving Parts | Best For | Limitations | Verification |
|---|---|---|---|---|---|
| Mag | Faraday's Law (Voltage) | None | Conductive fluids, slurries, wastewater. | Must be conductive; pipe must be full. | Bucket test; Ref meter. |
| Turbine | Rotor pulses | Yes | Clean liquids/gases; high accuracy. | Wear/debris; viscosity sensitive. | Bucket test; Pulse freq. |
| Paddle | Impeller pulses | Yes | Cooling systems; OEM; mid-accuracy. | Fouling-prone; insertion sensitive. | Bucket test. |
| Coriolis | Vibrating tubes | None | Mass flow; density; custody transfer. | Expensive; gas sensitive; heavy. | Ref meter; Check density. |
| Vortex | Shedding frequency | None | Liquids, gases, steam; wide temp. | Min flow req; low-flow cutoff. | DP/Temp calculation. |
| DP | flow = $K \times \sqrt{\Delta P}$ | None | Any fluid; low cost; well-understood. | Pressure loss; poor low-flow accuracy. | Measure DP directly. |

## MAGNETIC FLOW METERS

Mag meters are the workhorse of liquid flow measurement in process industries. No moving parts, minimal pressure drop, and they work with conductive liquids including water, chemicals, slurries, and wastewater. The one hard requirement: the pipe must be completely full. Air pockets and partial filling kill the reading.

| Failure | Symptom | Cause | Fix |
|---|---|---|---|
| Empty Pipe | Zero, erratic, or fluctuating reading. | Air pockets; meter at high point; pump suction air. | Relocate to low point or vertical "up" run. |
| Fouling | Gradual drift lower; erratic signal. | Scale, grease, or biological buildup on electrodes. | Clean electrodes; use ultrasonic/manual cleaning. |
| Grounding | Noisy signal; VFD interference. | Corroded grounding rings; stray pipe currents. | Verify grounding spec; clean/replace rings. |
| Coil Failure | Zero reading; fault code on display. | Wiring fault; failed electronics; blown fuse. | Check excitation voltage at coil terminals. |

*Table Magnetic Flow Meter Faults*

Shop-Floor Wisdom: "Mag meters lie when they're dry." Always confirm the pipe is full before chasing electronics or calibration.

## TURBINE FLOW METERS

A turbine meter has a rotor that spins as fluid flows through it. The rotation rate is proportional to flow velocity. A pickup coil or optical sensor detects each blade passage and outputs pulses—each pulse represents a fixed volume.

| Failure | Symptom | Cause | Fix |
|---|---|---|---|
| Fouled Rotor | Low reading; intermittent sticking; jammed rotor. | No upstream strainer; particulates or fibers in process. | Install/clean strainer; pull and clean meter internals. |
| Bearing Wear | Sluggish response; high minimum flow threshold. | Normal wear; poor lubrication; fluid contamination. | Replace bearings or internal cartridge. |
| Viscosity Shift | Drift with temperature; different products read differently. | Product change or temperature swing affecting Reynolds number. | Apply viscosity correction; recalibrate for current fluid. |
| Pickup Failure | No reading or erratic pulses while spinning. | Failed magnetic/optical pickup coil; broken wiring. | Verify pulses with frequency counter at the pickup. |

*Table Turbine Flow Meter Common Faults*

## CORIOLIS MASS FLOW METERS

Coriolis meters directly measure mass flow—not volume. Highly accurate and also measure density and temperature. Common in custody transfer, batching, and applications where mass rather than volume matters. They're expensive and sensitive to entrained gas, but nothing beats them for accuracy.

| Failure | Symptom | Cause | Fix |
|---|---|---|---|
| Entrained Gas | Erratic spikes; "Tube Imbalance" fault. | Air leaks; cavitation; gas coming out of solution. | Eliminate air ingress; move meter to high-pressure zone. |
| Vibration | Noisy readings; fixed offset errors. | Nearby pumps or compressors vibrating the piping. | Isolate with flex connections; add separate supports. |
| Coating/Erosion | Gradual drift; Density reading shifts. | Buildup on tube walls; abrasive slurry thinning the tubes. | Clean tubes; verify density against a known sample. |
| Electronics | No output; fixed reading; fault codes. | Failed transmitter; power supply instability. | Review diagnostics; check raw sensor drive gain. |

*Table Coriolis Flow Meter Common Faults*

## ULTRASONIC FLOW METERS

Ultrasonic flow meters measure flow velocity using sound waves transmitted through the pipe wall or through the fluid itself. Two types matter for field work:

Clamp-on ultrasonic: transducers strap to the outside of the pipe with no process penetration. No cutting, no welding, no process shutdown required. This makes clamp-on ultrasonics the most useful portable verification tool in flow measurement—bring one to any disputed reading and get an independent check in minutes.

In-line ultrasonic: transducers are wetted (inserted into the process). More accurate than clamp-on, used as permanent installations in custody transfer and utility metering.

Common failure modes for in-line and clamp-on:

Pipe coating or scale between the transducer and the pipe wall (clamp-on): signal cannot penetrate. Re-couple with fresh gel or relocate.

Entrained air or gas bubbles: disrupts signal propagation. Reading becomes erratic or drops out.

Incorrect pipe parameters entered: wall thickness, pipe OD, and pipe material affect the speed-of-sound calculation. Wrong parameters produce consistent but incorrect readings.

Transducer spacing wrong (clamp-on): spacing is calculated from pipe diameter and fluid type. The instrument calculates the correct spacing—use it.

💡 Pro Tip: A clamp-on ultrasonic meter belongs on every instrumentation technician's cart. When operators dispute a flow meter reading, a 10-minute clamp-on check either confirms the installed meter or isolates the problem to the meter itself—no process entry required.

## DIFFERENTIAL PRESSURE (DP) FLOW

An orifice plate, Venturi, or flow nozzle creates a restriction in the flow. Pressure drop across the restriction is proportional to the square of velocity. A DP transmitter measures the difference. Simple, robust, and still common—but the square root relationship reduces accuracy at low flows, and impulse lines are the same liability as in Chapter 15.

Key troubleshooting note: if the DP transmitter impulse lines plug on the high side or low side independently, the reading will be wrong in unpredictable ways. Always bleed and verify both legs.

## VORTEX FLOW METERS

A vortex meter detects the oscillating vortices shed by a bluff body (shedder bar) placed in the flow stream. Vortex frequency is proportional to flow velocity. No moving parts, good accuracy, and capable of measuring steam, gas, and liquid—which makes vortex meters the dominant technology for steam flow measurement in industrial plants.

Key troubleshooting notes: vortex meters have a minimum flow threshold (Reynolds number cutoff) below which the signal drops out entirely. At low flows the reading goes to zero—not because the meter has failed, but because flow is below the minimum detectable velocity. Piping vibration at frequencies near the vortex frequency causes false counts and elevated readings. Isolate the meter from structural vibration sources before chasing calibration. A meter that reads correctly at high flow but drifts or zeros out at low flow almost always has a minimum flow problem, not a hardware failure.

## COMMON FAILURE MODES ACROSS ALL TYPES

| Failure | Meters Affected | Symptom | Check First |
|---|---|---|---|
| No Power | All | Zero reading; dead transmitter display. | 24 VDC at terminals; fuses; breaker. |
| Wrong Scaling | All (Analog) | Reading in wrong range or units. | Raw counts vs. expected mA; PLC scaling. |
| Installation | All | Persistent inaccuracy; non-linear drift. | Upstream/downstream straight runs. |
| Signal Noise | All (Analog) | Reading fluctuates randomly. | VFD cable routing; shield grounding. |
| Valve Issue | All | Reading lower than expected. | Throttled upstream valves; debris in line. |

## TROUBLESHOOTING WORKFLOW

Follow the chain: Power → Signal → Process Conditions → Field Verification → Scaling

## Step 1: Verify Power and Signal

Is the transmitter powered? Is there a 4-20 mA signal (or pulse output)? Does the local display show a value?

## Step 2: Compare to Process Conditions

Is there actually flow? Check pump status, valve positions, and downstream demand. Does the reading change when you expect it to—valve opens, pump starts? A reading that doesn't respond to deliberate changes is either a dead signal path or a plugged impulse line (DP type).

## Step 3: Check Raw Counts and Scaling

Do raw counts match the mA signal? Is scaling correct for the meter's range? Is the totalizer accumulating correctly?

## Step 4: Bucket Test

For liquid flow, a bucket and stopwatch settle arguments faster than any calibration certificate.

| Step | Action | Notes |
|---|---|---|
| 1 | Get a container of known volume (e.g., 5-gallon bucket). | Verify volume with a measuring cup if uncertain. |
| 2 | Divert flow into the bucket. | Use a valve or hose; safety first with chemicals. |
| 3 | Start stopwatch at start; stop when full. | Use a helper for higher flow rates. |
| 4 | Calculate: *GPM* = Gallons/Time (min). | Ex: 5 gallons in 30 seconds = 5 gal ÷ 0.5min = 10 GPM. |
| 5 | Compare to meter reading. | ±2-3% is "Correct"; larger = investigate. |

*Flow Bucket Test*

## INSTALLATION CONSIDERATIONS

Flow meters are sensitive to installation. Turbulence from elbows, valves, and fittings causes significant errors—even on a perfectly good meter.

| Meter Type | Upstream (Min) | Downstream (Min) | Notes |
|---|---|---|---|
| Mag Meter | 5D | 3D | D = pipe diameter; more is always better. |
| Turbine | 10 | 5D | Critical after elbows; use conditioners if tight. |
| Coriolis | None | None | Major advantage; insensitive to profile. |
| Vortex | 15D | 5D | Most sensitive; dual-plane elbows need more. |
| DP (Orifice) | 20D | 5D | Per AGA/ISO standards for specific fittings. |

*Table Flow Meter Installation Considerations*

Orientation matters too. Mag meter electrodes should be horizontal—not at top (air) or bottom (sediment). Coriolis orientation varies by design—check manufacturer instructions. All meter types: avoid high points where air can collect.

## REAL-WORLD EXAMPLES

### Example 1: The Phantom Flow

**Problem**: Mag meter shows 50 GPM but the receiving tank level isn't increasing.

**Finding**: Pipe developed an underground leak. Flow was happening—straight into the ground. Meter was reading correctly; the product just wasn't reaching the tank.

**Lesson**: The meter tells you what's passing through it, not what reaches the destination. Check the whole system.

### Example 2: The Noisy Coriolis

**Problem**: Coriolis meter reading bounces ±15% randomly on a clean water system.

**Finding**: Small air bubbles entrained from a leaky pump seal. Bubbles too small to see but enough to disrupt tube vibration. Transmitter showed intermittent "tube amplitude" warnings.

**Fix**: Replaced pump mechanical seal. Readings stabilized immediately.

### Example 3: The Bucket Test Proves It

**Problem**: Operators insist the flow meter reads 20% low and want it recalibrated.

**Bucket test result**: 5 gallons in 28 seconds = 10.7 GPM. Meter reads 10.5 GPM. Difference: 2%—within the meter's ±1% accuracy specification.

**Lesson**: Don't recalibrate based on gut feel. Prove the meter is wrong before touching it. The bucket doesn't lie.

## DECISION TREE REFERENCE

A flow meter troubleshooting decision tree is available at BootsOnTheGroundTech.com for download and printing.

**Pro Tip:** For liquid flow, a bucket and stopwatch will settle arguments faster than any calibration certificate. If the bucket test agrees with the meter, the meter is right.

**Shop-Floor Wisdom:** "Mag meters lie when they're dry." Confirm the pipe is full before chasing electronics or calibration.

# CHAPTER 17
# LEVEL INSTRUMENTS

## LEVEL INSTRUMENTS

Level measurement tells you how much is in a tank, bin, or vessel—liquid, solid, or slurry. It's critical for inventory management, batch control, overflow prevention, pump protection, and safety interlocks.

Level instruments range from simple float switches to sophisticated radar transmitters. Each technology has its strengths, weaknesses, and failure modes. Matching the right technology to the application is half the battle. The other half is understanding why they fail and how to troubleshoot them.

## MEASUREMENT METHODS: POINT VS. CONTINUOUS

Point level detects when level reaches a specific point—high, low, or both. Output is discrete (ON/OFF). Used for alarms, interlocks, and simple control.

Continuous level measures actual level throughout the vessel range. Output is analog (4-20 mA) representing 0-100% of span. Used for inventory tracking, PID control, and trending.

Many applications use both: continuous measurement for control, with independent point-level switches for safety interlocks.

## SENSOR TYPES AT A GLANCE

| Type | Principle | Contact | Best For | Key Limitations |
|---|---|---|---|---|
| Float | Buoyant switch | Yes | Simple point level. | Moving parts stick; foam errors. |
| Ultrasonic | Sound pulse timing | No | Non-contact liquids/solids. | Foam absorbs sound; temp shifts. |
| Radar | Microwave timing | No | Extreme vapor/temp; high accuracy. | Expensive; low-dielectric issues. |
| GWR | Microwave on probe | Probe | Low-dielectric; turbulent tanks. | Probe coating/buildup; damage. |
| Capacitance | Dielectric change | Probe | High temp/pressure; slurries. | Affected by coating/moisture. |
| Hydrostatic | $P = h \cdot \rho \cdot g$ | Yes | Reliable, simple liquid level. | Requires fixed, known density. |

*Table Level Sensor Types*

## FLOAT SWITCHES

Simple and reliable for point-level detection. A buoyant float rises and falls with liquid, actuating a switch at a set point.

| Failure | Symptom | Cause | Test / Fix |
|---|---|---|---|
| Float stuck | Switch doesn't actuate even when level changes; stuck in one state | Product buildup or corrosion on float or stem; debris jamming float | Manually lift/depress float—if switch actuates, float was stuck; clean or replace |
| Switch failure | Float moves freely but no output change; intermittent output; welded contacts | Mechanical wear; corrosion; shock | With float in actuated position, check continuity through switch contacts |
| Lost buoyancy | Switch never actuates even at high level; float found at bottom | Float absorbed product over time; damaged float shell | Replace float; verify material compatibility with process fluid |

*Table Common Float Switch Faults*

## ULTRASONIC LEVEL SENSORS

Non-contact, top-mounted sensor that emits an ultrasonic pulse and times the echo from the surface. Simple to install and cost-effective, but sensitive to the environment inside the vessel.

# Level Instruments

| Failure | Symptom | Cause | Fix |
|---|---|---|---|
| Foam interference | Reading stuck at maximum (no echo detected); erratic tracking | Agitation creating foam; product characteristics (soap, chemicals, biological) | Some sensors have foam-rejection algorithms; consider radar for foam-prone apps |
| Condensation on transducer | Erratic readings; jumps or drops; worse after temperature changes | Moisture on transducer face interferes with pulse | Clean transducer face; use hydrophobic coating; improve ventilation |
| False echo from structure | Reading locks onto fixed distance; level jumps; no correlation | Agitators, baffles, inlet pipes, or welds reflecting the pulse | Aim sensor away from obstructions; use echo rejection / false echo mapping |
| Temperature effects | Reading drifts with ambient or vapor temperature | Speed of sound varies with temperature; no compensation enabled | Enable temperature compensation; some sensors have built-in probes |
| Dead zone | Reading maxes out or goes erratic when level rises close to sensor | All ultrasonic sensors have a near-field dead zone (inches to a foot) | Install sensor higher; choose sensor with shorter specified dead zone |

*Table Common Ultrasonic Level Sensor Faults*

## RADAR LEVEL SENSORS

Radar uses microwave energy instead of sound—less affected by foam, vapor, temperature, and dust than ultrasonic. The right answer for difficult applications where ultrasonic fails.

| Failure | Symptom | Cause | Fix |
|---|---|---|---|
| Antenna buildup | Reduced signal strength; erratic readings; loss of measurement | Product buildup on antenna absorbs or reflects signal | Clean antenna; consider sensor with purge connection or self-cleaning design |
| Low dielectric constant | Weak or lost signal; reading defaults to empty | Hydrocarbons, some plastics, and dry powders reflect radar weakly | Use guided-wave radar; select sensor designed for low-dielectric applications |
| Turbulence / agitation | Noisy fluctuating readings; signal loss during agitation | Violent surface motion scatters radar signal | Install in stilling well; increase signal averaging in transmitter |
| Multiple reflections | Reading shows level that doesn't exist; value is a multiple/fraction | In tall narrow vessels, radar can bounce multiple times | Use false-echo mapping; consider guided-wave radar |

*Table Common Radar Faults*

## HYDROSTATIC (PRESSURE-BASED) LEVEL

Pressure at the bottom of a liquid column is proportional to the height of liquid above it.

The physics: Pressure = Height × Density × Gravity. For water: 1 foot ≈ 0.433 PSI. If you measure 4.33 PSI at the bottom of a tank, you have approximately 10 feet of water.

| Failure | Symptom | Cause | Fix |
|---|---|---|---|
| Density not accounted for | Reading consistently high or low; accurate with water | Transmitter calibrated for water (SG = 1.0) but product is different | Apply correction: Actual Level = (Pressure ÷0.433)÷SG |
| Sensor fouling / plugging | Slow response; stuck reading; zero offset | Submersible sensor or pressure tap coated or plugged | Clean or replace sensor; clear tap or impulse line |
| Reference leg problem | Erratic readings; level reads wrong after maintenance | Wet leg drained or has air bubbles; dry leg has condensate | Refill wet leg; drain dry leg; verify reference conditions |
| Density change with temp | Reading drifts through the day with constant level | Product density changes as temperature changes | Add temperature compensation to the level calculation |

*Table Common Hydrostatic Level Measurement Faults*

## TROUBLESHOOTING WORKFLOW

Follow the chain: Physical Reference → Power → Signal → Environmental Check → Stimulate → Scaling

### Step 1: Visual Check—Know the Actual Level First

Before troubleshooting the instrument, verify actual level by any independent means available: sight glass, dipstick, opening a hatch and looking. You cannot troubleshoot a level instrument if you don't know what the true level is.

### Step 2: Check Power and Signal

Is the transmitter powered? Is there a 4-20 mA output (or relay state for switches)? Does the local display show a reading?

### Step 3: Compare to Actual

Sensor says 50%—is the tank actually half full? If the sensor matches reality, the problem is downstream (wiring, scaling, HMI).

## REAL-WORLD EXAMPLES

### Example 1: The Phantom High Level

**Problem**: High-level alarm on a water tank keeps triggering; operators say tank is only half full.

**Finding**: Foam on the water surface from recent chemical treatment. Ultrasonic sensor saw the foam as the surface and read 95%.

**Fix**: Increased sensor filtering. Long-term: changed to radar sensor, which sees through foam.

### Example 2: The Stuck Float

**Problem**: Low-level pump protection switch never trips; pump ran dry and burned up.

**Finding**: Product (syrup) had coated the float stem. Float was glued in place by dried product.

**Fix**: Replaced switch with cleaner design (external cage float). Added PM task to inspect quarterly.

### Example 3: The Drifting Hydrostatic Reading

**Problem**: Level reading drifts 5-10% over the course of a day with confirmed constant level.

**Finding**: Temperature of the product changed significantly from morning to afternoon, changing its density. Pressure changed even though level didn't.

**Fix**: Added temperature compensation to the level calculation.

### Example 4: The Bouncing Radar

**Problem**: Radar level reading bounces ±8% continuously in a mixing tank.

## Step 4: Environmental Check

| Technology | Environmental Factors to Check |
|---|---|
| Ultrasonic | Foam on surface; condensation on transducer; vapor or dust in vessel; false echoes from internal structure |
| Radar | Antenna buildup; turbulence or agitation; low-dielectric product; multiple reflection paths |
| Capacitance | Coating on probe; dielectric changes from temperature or product variation; conductive bridging |
| Hydrostatic | Density changes; plugged pressure tap; fouled submersible sensor; reference leg condition |
| Float switch | Float stuck or coated; debris in vessel; float lost buoyancy |

*Table Common Environmental Faults*

## Step 5: Stimulate or Simulate

**Float switch**: manually lift/lower the float. Ultrasonic sensors measure level by timing how long it takes for a sound pulse to travel from the sensor to the surface and back. They're non-contact—the sensor mounts at the top of the vessel and measures down to the liquid or solid surface.

**Ultrasonic/radar**: hold a reflector (bucket lid, board) at a known distance and verify reading changes correctly. Capacitance: lower probe into a water bucket for quick function check. Hydrostatic: apply known pressure with a hand pump. If the sensor responds correctly to stimulation but not to the process, the problem is environmental—not the sensor.

## Step 6: Check Scaling

If raw counts are correct but engineering units are wrong, it's a scaling problem in the PLC or HMI.

*Level Instruments*   147

**Finding**: Turbulent surface from aggressive agitation scattering the radar signal. Reading stable when agitator is off.

**Fix**: Installed a stilling well (vertical pipe open at top and bottom) to calm the surface in the measurement zone. Readings stabilized.

## DECISION TREE REFERENCE

A level instrument troubleshooting decision tree is available at BootsOnTheGroundTech.com for download and printing.

💡 Pro Tip: When a level sensor is giving you trouble, start with a sight glass or dipstick. Know the actual level before you start troubleshooting the instrument.

⚡ Shop-Floor Wisdom: If an ultrasonic level reading jumps around, check for foam or condensation before chasing wiring. Environment beats electronics most of the time.

# CHAPTER 18
# LOAD CELLS & SCALES

## LOAD CELLS & SCALES

Load cells measure weight or force. They're everywhere in manufacturing: batching systems, filling stations, check-weighers, hopper scales, truck scales, tension control, press-fit verification, and end-of-line quality checks.

A load cell converts mechanical force into an electrical signal—typically a millivolt output that gets amplified and conditioned into something the PLC can read. That signal chain is sensitive. Noise, wiring errors, mechanical binding, and environmental factors corrupt the measurement in ways that aren't immediately obvious.

This chapter covers how load cells work, how they fail, and how to troubleshoot them without wasting hours or replacing good parts.

## HOW LOAD CELLS WORK

Most industrial load cells use bonded strain gauges. When a metal element is stressed, its electrical resistance changes—increases under tension, decreases under compression. Four strain gauges are wired in a Wheatstone bridge: two in tension, two in compression. This arrangement doubles the output signal, cancels temperature effects, and provides a stable balanced output.

The load cell requires excitation voltage—typically 5 VDC or 10 VDC—applied across the bridge. Output is specified in millivolts per volt (mV/V) at full capacity. A 2 mV/V cell with 10V excitation outputs 20 mV at full rated load. At 50% load, it outputs 10 mV. This is a tiny signal—noise, wiring resistance, and connection quality matter enormously.

## LOAD CELL TYPES

| Type | Configuration | Best For | Notes | Limitations | Verification |
|---|---|---|---|---|---|
| Single-Point | Platform supported by one cell. | Bench scales; packaging. | Off-center loading is compensated. | Must be conductive/full pipe. | Bucket test. |
| Bending Beam | Perpendicular force (bending). | Hoppers; small platforms. | Sensitive to side loads. | Wear/viscosity sensitive. | Pulse freq. |
| Shear Beam | Shear stress measurement. | Tank & hopper weighing. | Industry standard; rugged. | Fouling/depth sensitive. | Bucket test. |
| Canister | Cylindrical compression. | Truck scales; heavy tanks. | Purely vertical; high capacity. | Expensive; gas sensitive. | Ref meter. |
| S-Type | S-shaped; Tension/Compression. | Hanging scales; in-line. | Easy to install in-line. | Min flow req; vibration. | DP calc. |
| Multi-Cell | 3 or 4 cells in a system. | Large vessels; floor scales. | Uses a summing box for signals. | Pressure loss; plugging. | Measure DP. |

*Table Common Load Cell Types*

## INDICATOR / SIGNAL CONDITIONER

The load cell's millivolt output requires an indicator or signal conditioner to provide excitation voltage, amplify the signal, filter noise, apply calibration (zero, span, linearization), and output a usable signal (4-20 mA, 0-10V, digital, or display).

Key calibration terms:

- Zero: Output at no load (tare weight of the empty platform or vessel)
- Span: Output change from zero to full capacity
- Linearity: How closely output tracks a straight line from zero to full scale

- Repeatability: How closely the same weight produces the same reading every time

## COMMON FAILURE MODES

| Failure | Symptom | Cause | Fix | Limitations | Verification |
|---|---|---|---|---|---|
| Binding | Hysteresis; no zero return. | Rigid piping; debris; rubbing. | Eliminate contact; use flex joints. | Must be conductive/full pipe. | Bucket test. |
| Cable Damage | Erratic readings; drift. | Forklifts; rodents; moisture. | Wiggle test; replace entire cable. | Wear/viscosity sensitive. | Pulse freq. |
| Moisture | Humidity drift; corrosion. | Failed seals; condensation. | Insulation test (>5000 MΩ); replace. | Fouling/depth sensitive. | Bucket test. |
| Overload | Permanent Zero Shift. | Shock loads; forklift impact. | Replace; metal is permanently deformed. | Expensive; gas sensitive. | Ref meter. |
| EMI Noise | Jittery reading. | VFD routing; dual-point ground. | Separate cables; 1-point shield. | Min flow req; vibration. | DP calc. |
| Summing Box | Corner errors; erratic data. | Corrosion; water; loose wires. | Inspect/dry; check each cell input. | Pressure loss; plugging. | Measure DP. |

*Table Common Load Cell Failures*

⚡ Shop-Floor Wisdom: Before you blame the load cell, push on the platform. Binding causes more scale problems than bad load cells.

## TROUBLESHOOTING WORKFLOW

Follow the chain: Mechanical → Excitation → Signal → Indicator → Scaling

### Step 1: Verify the Problem

Place a known weight on the scale. Does it read correctly? If it's within tolerance, the scale is OK—investigate elsewhere. If it's significantly off, proceed.

## Step 2: Check for Mechanical Binding

This is the most common cause of scale problems and takes 30 seconds to check.

| Check | What to Look For |
|---|---|
| Push Test | Push sideways; reading should not change significantly. |
| Clearance | Platform touching frames, guards, or structure. |
| Debris | Rocks, ice, or product buildup under the platform. |
| Flexibility | Rigid conduit or hard piping pulling on the vessel. |
| Check Rods | Seized or overtightened bumpers; they must be loose. |
| Hardware | Overtightened mounting bolts restricting flexure. |

*Table Common Load Cell Binding Faults*

## Step 3: Check Excitation Voltage

Measure excitation at the load cell terminals or junction box. Should be stable within ±0.1V of specification (typically 5 VDC or 10 VDC). Low or unstable excitation indicates a power supply or wiring problem.

## Step 4: Check Load Cell Output

Measure millivolt output at the load cell.

| Parameter | Example Value | Formula |
|---|---|---|
| Capacity | 1,000 lb | - |
| Sensitivity | 2 mV/V | - |
| Excitation | 10 VDC | - |
| Full-Scale | 20 mV | Sensitivity × Excitation |
| At 50% Load | 10mV | (Load ÷ Capacity) × Full-Scale |
| At Zero | ≅ 0 mV | Within ±0.1 mV of zero |

*Table Load Cell Output Check*

If output doesn't match expectation for the applied load, the load cell is suspect.

## Step 5: Wiggle Test

With the scale at rest, wiggle and flex the load cell cable, junction box connections, and cable runs. Any change in reading indicates a wiring problem at the location being flexed.

## Step 6: Insulation Resistance Test

With load cell disconnected, measure insulation resistance from each conductor to ground. Should be >5,000 M$\Omega$. Low readings indicate moisture or insulation damage.

## Step 7: Check the Indicator

If load cells check out OK, verify indicator calibration (zero, span), check 4-20 mA output to PLC, and verify PLC scaling matches indicator output range.

# STRAIN GAUGE RESISTANCE: THE BASELINE CHECK

Before connecting a load cell to an indicator or performing any electrical measurement, know what a healthy load cell measures. Most industrial load cells use a 350-ohm or 700-ohm Wheatstone bridge. With the load cell completely disconnected from the indicator, measure resistance across the excitation terminals (EXC+ to EXC−) with a standard ohmmeter. You should read the rated bridge resistance (±5%). Measure signal output terminals (SIG+ to SIG−)—the same value.

If you read open circuit (infinite resistance): the bridge is broken. At least one strain gauge has failed, or a conductor inside the load cell cable is open.

If you read near zero: the bridge is shorted. Overload damage or moisture ingress is the most common cause.

If you read the wrong value: partial damage to one gauge, or a wiring error at the junction box.

This 30-second check tells you whether a load cell is worth testing further or ready to be replaced before you touch the indicator or calibration procedure.

## CALIBRATION BASICS

Calibration should follow, not replace, troubleshooting. If a scale needed a huge zero or span adjustment to read correctly, something is wrong—find and fix the root cause (binding, damaged cell, wiring) before recalibrating.

| Step | Procedure | Notes |
|---|---|---|
| 1. Zero | Clear scale; set display to zero. | Always do this first. |
| 2. Span | Apply known test weight (near max). | Use certified weights for legal-for-trade. |
| 3. Linearity | Check weights at 25/50/75%. | If it's not linear, the cell is damaged. |

*Table Load Cell Calibration Basics*

## REAL-WORLD EXAMPLES

### Example 1: The Mystery Zero Shift

**Problem**: Hopper scale reads 200 lb empty. Was reading zero last month.

**Finding**: Flexible conduit connecting the weighed hopper to the structure had become rigid with age and cold weather. Conduit was bridging the mechanical gap and transmitting force.

**Fix**: Replaced rigid section with new flexible conduit. Zero returned to normal immediately.

## Example 2: The Jumpy Batching Scale

**Problem**: Ingredient batching scale reading jumps ±5 lb randomly during filling.

**Finding**: Load cell cable running in the same tray as VFD output wiring on a nearby conveyor. Electrical noise from VFD coupling into the load cell signal.

**Fix**: Rerouted load cell cable in a separate tray, minimum 18" from power wiring. Added shielded cable. Noise eliminated.

## EXAMPLE 3: THE CORNER THAT DIDN'T WORK

**Problem**: Four-cell platform scale reads 5% low when weight is placed on one corner.

**Finding**: Water had entered one load cell through a damaged cable gland. Corrosion on the strain gauges.

**Fixed**: Replaced the load cell and cable. Sealed all glands properly. Added inspection to PM program.

## Example 4: The Forklift Lesson

**Problem**: Floor scale reads 500 lb at zero after being struck by a forklift.

**Finding**: Overload damage from impact. One load cell plastically deformed—permanent zero offset.

**Fix**: Replaced load cell. Installed protective bollards around the scale perimeter.

## QUICK FIELD VERIFICATION

| Method | Procedure | What It Proves |
|---|---|---|
| Known Weight | Place a verified weight on scale. | General accuracy. |
| Substitution | Swap a suspect cell with a good one. | Proves if fault is the Cell or the Wiring. |
| Push Test | Lateral force on platform. | Detects mechanical binding. |

*Table Load Cell Field Verification*

## DECISION TREE REFERENCE

A load cell troubleshooting decision tree is available at BootsOnThe-GroundTech.com for download and printing.

💡 Pro Tip: Before replacing a load cell, verify the mechanical installation is free and clear. Binding causes more scale problems than bad load cells.

⚡ Shop-Floor Wisdom: If you can change the reading by pushing sideways on the platform, something is binding. Fix the mechanics before blaming the electronics.

# CHAPTER 19
# PHOTOELECTRIC SENSORS

Photoelectric sensors—photoeyes—use light to detect the presence, absence, or position of objects. They're the eyes of discrete manufacturing: detecting parts on conveyors, verifying box presence before labeling, confirming bottles before filling, counting products, and triggering downstream operations.

When a photoeye fails or misbehaves, the line stops, parts pile up, or products get missed. Most photoeye problems fall into a handful of categories—and once you know what to look for, troubleshooting is fast.

## SENSOR TYPES AT A GLANCE

| Type | Sensing Method | Range | Strengths | Best Application |
|---|---|---|---|---|
| **Through-Beam** | Separate emitter/receiver. | 100+ ft | Most reliable; cuts through dirt/dust. | Safety curtains; long distances. |
| **Retroreflective** | Bounces off a reflector. | ~30 ft | One device to wire; strong signal. | Conveyors; non-shiny objects. |
| **Diffuse** | Reflects off the target. | ~3 ft | Simplest setup; no reflector needed. | Presence detection; close range. |
| **BGS** | Triangulation (proximity). | ~3 ft | Ignores background surfaces/colors. | Dark parts on light backgrounds. |
| **Laser** | Focused tight beam. | Varies | Precise positioning; tiny spot size. | Small parts; edge detection. |

*Table Photoelectric Types*

See Appendix G for the complete sensor selection matrix.

## LIGHT-OPERATE VS. DARK-OPERATE

This is the most common source of hidden failures in photoeye installations—and it's entirely a configuration issue, not a hardware failure.

Light-Operate (L.O.): Output is ON when the receiver sees light. Output is OFF when light is blocked.

Dark-Operate (D.O.): Output is ON when light is blocked. Output is OFF when the receiver sees light.

The classic mistake: sensor LED lights up when the part is present. Technician assumes it's working. But the PLC never sees the part. Why? The sensor is set to Light-Operate on a through-beam. Light-Operate means output is HIGH when light is received—which for a through-beam means when the beam is NOT blocked. The LED behavior was correct; the mode was wrong.

Always verify three things:

1 What does the sensor LED indicate? (LED ON usually = output ON)

2 What does the PLC expect? (HIGH when part present, or LOW?)

3 Does the sensor mode match the PLC expectation?

Many sensors have a switch or teach button to change modes. Check it before condemning the sensor.

## OUTPUT TYPES: PNP VS. NPN

Photoelectric sensors have transistor outputs—identical in concept to proximity sensors and discrete inputs covered in earlier chapters.

PNP (Sourcing): Output goes HIGH (+24V) when active. Common in North America and Europe. Works with sinking input cards.

NPN (Sinking): Output goes LOW (0V) when active. Common in Asia and some OEM equipment. Works with sourcing input cards.

The sensor output type must match the PLC input card type. Mismatch = no detection or always-on detection. See Chapter 11 (Discrete Inputs) for full PNP/NPN troubleshooting detail.

## COMMON FAILURE MODES

| Failure | Symptom | Cause | Fix |
|---|---|---|---|
| Dirty Lens | Intermittent detection; range drop. | Dust, oil mist, or residue buildup. | Clean lens; use an air-purge hood. |
| Misalignment | Intermittent signal; low gain. | Vibration or bumps shifting mounts. | Realign and tighten hardware. |
| L.O. / D.O. | Logic is inverted in the PLC. | Light-On/Dark-On switch set wrong. | Correct sensor mode switch. |
| PNP/NPN | LED works; PLC input is dead. | Sensor output type mismatch. | Match sensor to Input Card type. |
| Shiny Target | Misses foil or polished metal. | Target acts like a reflector. | Use a Polarized retroreflective sensor. |
| Background | Triggers with no part present. | Background reflects light. | Use Background Suppression (BGS). |
| Ambient Light | Fails in direct sunlight. | Receiver saturation from sun/lights. | Add a sun shield or hood. |

*Table Common Photoelectric Failure Modes*

⚡ **Shop-Floor Wisdom**: If the conveyor "stops detecting" parts, clean the photoeye before you touch the PLC. Nine times out of ten, that's the fix.

Photoelectric sensors are among the most reliable devices in industrial automation—until something gets in the way of the beam. Dust, mist, vibration, misalignment, and shiny targets cause 90% of photoeye problems. The sensor itself rarely fails. When troubleshooting, verify the optical path first, check alignment second, and suspect the sensor last.

## TROUBLESHOOTING WORKFLOW

Follow the chain: LED → Lens → Power → Output → Mode → Alignment → Environment

### Step 1: Observe the Sensor LED

Does the LED change state when the target is presented and removed? LED changes correctly → sensor is detecting; problem is downstream (wiring, PLC input, logic). LED doesn't change → sensor isn't detecting; check alignment, clean lens, verify power. LED always ON or always OFF → check mode (L.O./D.O.), target presence, or sensor failure.

### Step 2: Clean the Lens

Before doing anything else, wipe the lens—and the reflector if applicable. A remarkable number of "sensor failures" are just dirty lenses.

### Step 3: Check Power

Verify supply voltage at the sensor: Brown wire = +24 VDC, Blue wire = 0 VDC. No power = no detection.

### Step 4: Meter the Output

With sensor detecting (LED ON): PNP sensor output (black wire) should be ~+24V. NPN sensor output (black wire) should be ~0V. Compare to what the PLC input expects.

### Step 5: Jumper Test

Disconnect the sensor and jumper the PLC input to simulate detection. If the PLC sees the jumper, the problem is in the sensor or wiring to the sensor. If the PLC doesn't see the jumper, the problem is the PLC input card, slot, or configuration.

## Step 6: Check Alignment (Through-Beam and Retroreflective)

Is the emitter aimed at the receiver or reflector? Is the reflector clean and undamaged? Has anything vibrated loose? Use the signal strength indicator if available—maximize signal through alignment.

## Step 7: Check for Environmental Factors

Shiny target fooling a retroreflective sensor? Background triggering a diffuse sensor? Ambient light (sunlight, welding arc, strobe) interfering? Vibration causing intermittent detection?

## Step 8: Check Mode and Output Type

Light-Operate vs. Dark-Operate—does it match PLC logic? PNP vs. NPN—does it match the input card?

# REAL-WORLD EXAMPLES

## Example 1: The Dirty Reflector

**Problem**: Retroreflective sensor intermittently misses boxes on a conveyor; gets worse as the shift goes on.

**Finding**: Reflector covered in cardboard dust. Cleaned reflector; detection restored. Added weekly reflector cleaning to PM checklist.

### Example 2: The Shiny Shrink-Wrap

**Problem**: Retroreflective sensor misses shrink-wrapped packages but detects cardboard boxes reliably.

**Finding**: Shrink wrap reflecting light back to sensor—standard retroreflective sensor cannot distinguish the product reflection from the actual reflector.

**Fix**: Replaced with polarized retroreflective sensor and polarized reflector. Shiny surface no longer fools the sensor.

### Example 3: The Backwards Logic

**Problem**: Photoeye detects parts (LED lights) but the filling machine acts like no bottle is present.

**Finding**: Through-beam sensor set to Light-Operate. Output goes ON when beam is received (no part present) and OFF when beam is blocked (part present). PLC expected HIGH when part was there.

**Fix**: Changed sensor to Dark-Operate. Output now goes HIGH when beam is blocked. Machine works correctly.

### Example 4: The Afternoon Mystery

**Problem**: Sensor works perfectly until mid-afternoon, then starts missing parts randomly.

**Finding**: Diffuse sensor on a conveyor near west-facing windows. Afternoon sun shining directly into the receiver, saturating it and preventing detection of the reflected signal from the target.

**Fix**: Added a simple sheet metal sun shield above the sensor.

## ENVIRONMENTAL CONSIDERATIONS

| Environment | Recommendation |
|---|---|
| Dust and Dirt | Use through-beam sensors; schedule PM lens cleaning; use air purge ports. |
| Oil / Coolant | Use IP69K rated sensors; position away from direct spray. |
| Ambient Light | Use modulated sensors; install sheet-metal sun shields. |
| Temperature | Monitor for condensation in cold zones; check ratings for high-heat areas. |
| Vibration | Use robust mounts with lock washers; periodically verify alignment. |
| Hardware | Overtightened mounting bolts restricting flexure. |

*Table Photoelectric Environmental Fault Considerations*

## DECISION TREE REFERENCE

A photoelectric sensor troubleshooting decision tree is available at Boots-OnTheGroundTech.com for download and printing.

Pro Tip: Before troubleshooting wiring or PLC logic, clean the lens and reflector. It takes 10 seconds and solves most photoeye problems.

Shop-Floor Wisdom: If the LED changes correctly but the PLC doesn't see it, check Light-Operate vs. Dark-Operate first. Mode mismatch is the classic hidden failure.

# CHAPTER 20
# PROXIMITY SENSORS

Proximity sensors detect the presence of objects without physical contact. They're the workhorses of discrete manufacturing—detecting parts in fixtures, confirming cylinder positions, sensing metal targets on conveyors, verifying tool presence, and triggering countless operations on every automated line.

Unlike photoeyes, proximity sensors don't use light. They detect targets using electromagnetic fields (inductive), electric fields (capacitive), or magnetic fields (reed switch and Hall effect). Understanding which technology you're dealing with is the first step in troubleshooting.

## SENSOR TYPES AT A GLANCE

| Type | Detection Principle | Detects | Typical Range | Strengths |
|---|---|---|---|---|
| Inductive | Electromagnetic field. | Metal only. | 1–30 mm | Rugged; fast; immune to oil/dirt. |
| Capacitive | Electrostatic field. | Metal & Non-metal. | 1–30 mm | Detects through tank walls (liquids). |
| Reed Switch | Magnet closes contacts. | Magnets only. | Proximity-based | Simple; no power required. |
| Hall Effect | Voltage from magnetic field. | Magnets only. | Proximity-based | No moving parts; vibration resistant. |

*Table Proximity Types*

Each type has specific applications where it excels. Inductive sensors dominate metal detection in harsh environments. Capacitive sensors handle non-metallic targets and liquids. Reed switches and Hall Effect sensors work where no power is available at the sensor. Choosing the wrong type for your application guarantees detection problems—match the technology to the target material first.

## INDUCTIVE PROXIMITY SENSORS

| Target Material | Detection Distance vs. Steel | Notes |
|---|---|---|
| Mild Steel | 100% (Rated) | The baseline for all datasheet specs. |
| Stainless (304) | 60–90% | Varies by grade; non-ferrous grades are lower. |
| Aluminum | 40–50% | Significantly reduces range. |
| Brass / Copper | 25–45% | The "worst" targets for standard inductive. |
| Cast Iron | 95–100% | Nearly identical to mild steel. |

*Table Inductive Proximity Sensors*

💡 Field Tip: If a sensor has a rated sensing distance of 10 mm for steel, it will only detect aluminum at 4-5 mm. Set actual operating distance to 70-80% of rated maximum for reliable detection.

## FLUSH VS. NON-FLUSH MOUNTING

## CAPACITIVE PROXIMITY SENSORS

Inductive sensors detect metal. Capacitive sensors detect anything that disturbs an electrostatic field—plastic parts, liquids, granules, powders, wood, glass, and foam. They are the right answer when the target is non-metallic or when you need to detect liquid level through a non-metallic tank wall.

How they work: the sensor face generates an electrostatic field. When a target enters the field and changes its capacitance by a sufficient amount, the output switches. Most capacitive sensors have a sensitivity (gain) adjustment—a potentiometer or teach button—that sets the detection threshold.

## Key differences from inductive sensors:

Sensitivity adjustment is required. Unlike an inductive sensor that works at a set distance for a given metal, a capacitive sensor must be tuned to the specific target material. Too high a sensitivity setting and the sensor triggers on air movement or condensation. Too low and it misses the target entirely.

Non-metallic targets have different dielectric constants. Water and aqueous solutions (high dielectric) are easier to detect than dry granules or powders (low dielectric). Sensor range specs are typically given for a grounded metal plate—actual range for plastic or liquid targets is 50-80% of that rating.

## Common failure modes:

Condensation or product buildup on the sensor face triggers false detects—more common with capacitive sensors than inductive because the sensor responds to any dielectric change. Clean the face and reduce sensitivity before replacing.

**Target material change**: if the product changes (different granule size, different liquid density, different container material), a previously working sensor may need to be re-taught.

**Sensitivity drift**: some capacitive sensors drift over temperature. If a sensor works in the morning but false-triggers or misses targets as the machine heats up, check the installation temperature against the sensor's rated operating range.

**Troubleshooting workflow**: follow the same chain as inductive—LED, face, power, distance, output, jumper test—with one addition: after cleaning and power verification, adjust the sensitivity potentiometer across its full range while the target is present. If the LED never activates across the full sensitivity range, the sensor has failed or the target material is incompatible with the sensor type.

. . .

💡 Pro Tip: Capacitive sensors can detect liquid through a plastic tank wall. If you need a non-invasive level switch on a plastic vessel, a capacitive sensor mounted externally and tuned to the liquid is faster and cheaper than cutting a port.

| Type | Sensing Field | Mounting Requirement | Common Mistake |
|---|---|---|---|
| Flush (Shielded) | Focused forward. | Can be recessed into metal. | Using non-flush where flush is needed. |
| Non-Flush | Wide side-field. | Needs 2-3X diameter clearance. | Mounting "flush" in a steel bracket. |
| Lost buoyancy | Switch never actuates even at high level; float found at bottom | Float absorbed product over time; damaged float shell | Replace float; verify material compatibility with process fluid |

*Table Flush vs non-Flush*

## OUTPUT TYPES: PNP VS. NPN

Identical in concept to Chapter 11 (Discrete Inputs) and Chapter 19 (Photoelectric Sensors)—covered in full detail there. The key reminder for proximity sensors:

Normally Open (NO): Output is OFF with no target present; turns ON when target is detected.

Normally Closed (NC): Output is ON with no target present; turns OFF when target is detected.

Many sensors offer both NO and NC, or a switch selection. Always verify the replacement sensor matches the original in output type (PNP/NPN) and normal state (NO/NC) before installation.

# COMMON FAILURE MODES

| Failure | Symptom | Cause | Fix |
|---|---|---|---|
| Metal Buildup | LED always ON. | Chips or swarf covering the face. | Clean face; use an air-purge. |
| Material Error | Inconsistent detection. | Aluminum target/wrong distance. | Apply correction factor; move closer. |
| Mounting Error | LED always ON. | Non-flush sensor in metal bracket. | Replace with Flush-mount sensor. |
| PNP/NPN | LED OK; PLC dead. | Polarity mismatch with input card. | Match sensor to PLC Input type. |
| Overheating | Range drops in PM. | Mounted near oven/heat source. | Add heat shield; use High-Temp model. |
| Target Size | Inconsistent triggers. | Target is smaller than "Standard." | Move closer; use smaller sensor. |
| Ambient Light | Fails in direct sunlight. | Receiver saturation from sun/lights. | Add a sun shield or hood. |

*Table Proximity Common Failure Modes*

# TROUBLESHOOTING WORKFLOW

Follow the chain: LED → Face → Power → Distance → Mounting → Output → Jumper → Wiggle → Substitution

## Step 1: Observe the Sensor LED

LED changes correctly when target enters and leaves → sensor is detecting; problem is downstream (wiring, PLC, logic). LED always ON → detecting something—metal chips on face, mounting interference, or target stuck in range. LED always OFF → not detecting; check target distance, clean face, verify power.

## Step 2: Clean the Sensor Face

Wipe off any metal chips, oil, or debris. Retest before doing anything else.

### Step 3: Check Power

Brown wire = +24 VDC, Blue wire = 0 VDC. No power = no detection.

### Step 4: Check Target Distance and Material

Is the target within rated range for the actual target material? Operating distance should be 70-80% of rated maximum for reliable detection. Apply correction factor for non-steel targets.

### Step 5: Check Mounting

Flush sensor: can be flush with metal. Non-flush sensor: needs clearance from surrounding metal equal to 2-3× sensor diameter. If non-flush is mounted flush, surrounding metal is interfering.

### Step 6: Meter the Output

With sensor detecting (LED ON): PNP black wire should be ~+24V; NPN black wire should be ~0V. If voltage is correct but PLC doesn't see it, the problem is the PLC input card or configuration.

### Step 7: Jumper Test

Disconnect sensor and jumper the PLC input. If PLC sees the jumper, the problem is the sensor or wiring to it. If PLC doesn't see the jumper, the problem is the input card, slot, or configuration.

### Step 8: Wiggle Test

Manipulate the sensor cable while watching the LED and PLC input. Any intermittent behavior indicates cable damage.

## Step 9: Substitution Test

Swap with a known good sensor of the same type. Does the problem follow the sensor or stay with the location? Follows sensor = bad sensor. Stays with location = bad wiring, mounting problem, or mechanical interference.

## PNEUMATIC CYLINDER POSITION SENSING

Most pneumatic cylinders have a magnet embedded in the piston. Magnetic sensors (reed switch or Hall effect) mount on the outside of the cylinder barrel in T-slots or grooves and trigger when the piston magnet reaches the sensor position.

| Problem | Symptom | Fix |
|---|---|---|
| Alignment | Never triggers. | Slide sensor along slot until it catches. |
| Weak Magnet | Intermittent detection. | Replace cylinder; use higher sensitivity. |
| Brand Mismatch | Won't detect. | Use sensors matched to cylinder model. |
| Tie-Rod | Varying detection. | Position away from steel tie-rods. |
| Cast Iron | 95–100% | Nearly identical to mild steel. |

*Table Pneumatic Cylinder Positioning*

## REAL-WORLD EXAMPLES

### Example 1: The Phantom Part

**Problem**: Inductive sensor always shows "part present" even with fixture empty.

**Finding**: Non-flush (unshielded) sensor mounted flush in a steel bracket. The bracket itself was being detected.

**Fix**: Replaced with flush-mount (shielded) sensor of the same size. Problem solved.

## Example 2: The Inconsistent Detection

**Problem**: Proximity sensor detects most parts but misses some randomly—all parts the same shape and size.

**Finding**: Missed parts were aluminum; detected parts were steel. Sensor rated for 8 mm on steel. Aluminum parts only detectable at ~3.5 mm—some parts were positioned slightly farther away.

**Fix**: Moved sensor closer. Adjusted fixture to ensure consistent part position within 3 mm of sensor face.

## Example 3: The Chip Magnet

**Problem**: Cylinder-extended sensor in a CNC machine constantly reads "extended" even with cylinder retracted.

**Finding**: Sensor face covered with fine steel chips carried by flood coolant.

**Fix**: Cleaned sensor face. Installed air purge line to blow chips off the sensor. Added to daily PM checklist.

## Example 4: The Wrong Wire

**Problem**: Replaced a failed proximity sensor with "the same thing" from the storeroom. PLC never sees the part.

**Finding**: Original sensor was PNP; replacement was NPN. Storeroom had both versions—wrong one was grabbed.

**Fix**: Installed correct PNP sensor. Labeled storeroom bins to distinguish PNP from NPN.

## DECISION TREE REFERENCE

A proximity sensor troubleshooting decision tree is available at BootsOnTheGroundTech.com for download and printing.

💡 Pro Tip: When an inductive sensor acts up, clean the face first. Metal chips are the #1 cause of false triggers and missed detections.

⚡ Shop-Floor Wisdom: If the sensor LED works right but the PLC doesn't see it, check PNP vs. NPN before you replace anything. The storeroom might have both types—and they look identical.

# CHAPTER 21
# CABLE ASSEMBLIES

Cables are the nervous system of industrial automation. Every signal from every sensor, every command to every actuator travels through cables. And cables fail—more often than most people realize.

I've seen technicians replace three sensors, a PLC input card, and call in an integrator before someone finally wiggled a cable and found the intermittent open. Cables are unglamorous. They're not as interesting as PLCs or sensors. But they're responsible for a huge percentage of field failures.

This chapter covers the cables and connectors you'll encounter, how they fail, and how to find those failures without wasting hours.

## CABLE TYPES

| Material | Properties | Best For | Avoid For |
|---|---|---|---|
| PVC | Low cost; general purpose. | Fixed, dry installations. | Repeated flexing (cracks). |
| PUR | High flex life; oil resistant. | Cable carriers; robots. | No major industrial drawbacks. |
| TPE | Halogen-free; flame retardant. | Food & Pharma apps. | Heavy oil contamination. |
| Silicone | Extreme temp range. | Near ovens and furnaces. | Oils and cutting fluids. |
| PTFE | Extreme chemical resistance. | Process fluid contact. | High cost; stiff. |
| Target Size | Inconsistent triggers. | Target is smaller than "Standard." | Move closer; use smaller sensor. |
| Ambient Light | Fails in direct sunlight. | Receiver saturation from sun/lights. | Add a sun shield or hood. |

*Table Cable Types*

Rule: If the cable is in a cable carrier, robot dress pack, or any moving application, use PUR or a flex-rated cable. PVC will crack and fail. This is the single most common cable installation error in manufacturing.

Shielded vs. Unshielded: Use shielded cable for all analog signals (4-20 mA, 0-10V), long cable runs, cables routed near VFDs or power wiring, and any application where noise is a concern. Ground the shield at ONE end only—typically at the PLC or control panel. Grounding both ends creates a ground loop that worsens noise.

## CONNECTOR TYPES

| Coding | Pins | Primary Application | Notes |
|---|---|---|---|
| A-coded | 3, 4, 5, 8 | Sensors, Actuators, I/O. | Most common; 4-pin is standard. |
| B-coded | 5 | Legacy Fieldbus (Profibus). | Being phased out. |
| D-coded | 4 | Industrial Ethernet (100 Mbps). | Used for standard IO-Link/Ethernet. |
| X-coded | 8 | High-speed Ethernet (10 Gbps). | Physically larger; vision systems. |
| PTFE | Extreme chemical resistance. | Process fluid contact. | High cost; stiff. |
| Target Size | Inconsistent triggers. | Target is smaller than "Standard." | Move closer; use smaller sensor. |
| Ambient Light | Fails in direct sunlight. | Receiver saturation from sun/lights. | Add a sun shield or hood. |

*Table M12 Cable Assembly*

Note: This is the most common convention but not universal. Always verify with the sensor documentation. Some manufacturers use different color codes.

Male vs. Female: Sensors typically have a male (pins) connector. Cables typically have a female (sockets) connector on the sensor end. Straight vs. Right-Angle: Right-angle connectors save space in tight installations but are more prone to damage from side impacts.

## CABLE FAILURE MODES

Cable failures rarely happen instantly. They develop over time through mechanical stress, environmental exposure, or poor installation. The table below shows common failure modes and how to prevent them.

Recognizing failure patterns saves time. If you see intermittent signals, think flex fatigue or connector issues. If you see shorted or severed wires, look for mechanical damage at pinch points. If you see corrosion, trace the water source.

| Failure | Symptom | Cause | Prevention |
|---|---|---|---|
| Flex Fatigue | Intermittent signal. | Repeated bending; tight radius. | Use PUR flex-rated cable. |
| Mech. Damage | Severed or shorted wires. | Forklifts; pinch points. | Use conduit or armored cable. |
| Connector | High resistance; open loop. | Rough handling; cross-threading. | Align keyways; hand-tighten. |
| Water Ingress | Leakage currents; corrosion. | Damaged jacket; poor seals. | Use IP69K; route to drain water. |
| Chemical | Swelling/cracking jacket. | Cutting oils; acids; solvents. | Verify jacket compatibility. |
| Rodents | Visible chew marks. | Food plants; warehouses. | Use armored cable or conduit. |
| Ambient Light | Fails in direct sunlight. | Receiver saturation from sun/lights. | Add a sun shield or hood. |

*Table Cable Failure Modes*

When you suspect cable damage, systematic testing isolates the problem faster than visual inspection alone.

## TROUBLESHOOTING CABLE PROBLEMS

The five tests below isolate cable faults systematically. Start with the wiggle test—it's fast and finds most intermittent problems. Move to continuity and insulation tests only if the wiggle test shows nothing. The substitution test is your final proof.

| Test | Procedure | What You Find |
|---|---|---|
| Wiggle Test | Flex cable while watching PLC. | Pinpoints the exact break location. |
| Continuity | Check end-to-end resistance. | Finds open conductors or shorts. |
| Tug Test | Gently pull cable at connector. | Detects failed strain relief/terminations. |
| Insulation | Megohmmeter test (Conductor to Gnd). | Finds moisture or insulation breakdown. |
| Substitution | Swap with a known good cordset. | Quickly proves the cable is the "liar." |

*Table Cable Troubleshooting*

## CONNECTOR CARE

Most "intermittent" sensor problems trace back to connector issues—corrosion, bent pins, loose coupling nuts, or contamination. Proper connector maintenance prevents 90% of these failures.

| Action | Correct Practice | Common Mistake to Avoid |
|---|---|---|
| Mating | Align keyway; push straight in; tighten coupling nut hand-tight; turn the nut, not the cable. | Forcing misaligned connector—bends pins; cross-threading; overtightening; rotating cable body. |
| Unmating | Loosen coupling nut first; pull straight out. | Yanking cable without loosening nut; pulling at an angle. |
| Storage | Install dust caps on all unused connectors and receptacles. | Leaving connectors open—dirt and moisture cause high resistance and corrosion. |
| Cleaning | Use electronics-grade contact cleaner; lint-free swab for sockets; dry completely. | Using compressed air alone (blows debris deeper); using solvents that attack seals. |
| Inspection | Look for bent pins, corrosion (green/white), cracked housing, or damaged O-rings. | Assuming a connector is good because it looks OK from the outside—internal damage is hidden. |

*Table Connector Mating Maintenance*

## CABLE ROUTING BEST PRACTICES

| Guideline | Requirement | Why It Matters |
|---|---|---|
| Separation | 12" from power cables. | Prevents induced EMI/VFD noise. |
| Fixed Bend | Minimum 4 × D. | Prevents internal conductor stress. |
| Flex Bend | Minimum 10 × D. | Avoids premature flex fatigue. |
| Strain Relief | Every termination point. | Prevents tension on electrical pins. |
| Service Loops | Leave slack at both ends. | Allows for machine motion and repairs. |

*Table Cable Routing Best Practices*

How you route cables matters as much as the cable itself. Poor routing creates mechanical stress, picks up electrical noise, and makes troubleshooting harder. The guidelines below prevent the most common installation mistakes—violations that cause failures months or years after commissioning when everyone has forgotten how the cable was installed.

## REAL-WORLD EXAMPLES

### Example 1: The Phantom Sensor Failure

**Problem**: Proximity sensor intermittently stops detecting. Replaced sensor twice—problem persists.

**Finding**: Wiggle test at the sensor head caused dropouts. Flex fatigue at the cable exit point. Conductor broken internally.

**Fix**: Replaced cable with flex-rated PUR cordset. Adjusted cable routing to reduce flexing at sensor head. Added strain relief.

### Example 2: The Wet Connector

**Problem**: Photoeye on an outdoor conveyor works fine on dry days, fails randomly when it rains.

**Finding**: O-ring on the cable connector was missing—fell out during a previous service call and wasn't noticed. Water wicking into connector.

**Fix**: Replaced cordset. Verified O-ring present. Tested seal.

## Example 3: The Cable Carrier Casualty

**Problem**: Multiple sensors on a linear axis fail every 2-3 months.

**Finding**: Cables in the carrier were PVC-jacketed. PVC is not flex-rated. The constant motion was fatiguing the conductors.

**Fix**: Replaced all cables in the carrier with PUR flex-rated cordsets. Verified bend radius at carrier ends. No further failures.

## Example 4: The Forklift Strike

**Problem**: Entire zone of I/O stops working after night shift.

**Finding**: Forklift ran over floor-level conduit and severed the trunk cable inside.

**Fix**: Replaced cable. Relocated conduit to a protected overhead path. Added bollards to protect the new route.

## Example 5: The Storeroom Surprise

**Problem**: New sensor installed; doesn't work. Sensor tests fine on bench.

**Finding**: Cable from storeroom was a 3-pin cordset. Sensor requires 4-pin. The pin functions are different between the two.

**Fix**: Installed correct 4-pin cordset. Labeled storeroom bins more clearly.

## SPARES TO KEEP ON HAND

| Item | Why |
|---|---|
| M12 4-pin Cordsets | Straight and Right-Angle; covers 80% of sensors. |
| M12 5-pin Cordsets | For dual-output or shielded sensors. |
| M8 4-pin Cordsets | For miniature proximity and photo-eyes. |
| M12 Field-Wireables | For emergency repairs of custom lengths. |
| M12 D-coded | For Industrial Ethernet/IO-Link repairs. |
| Panel Receptacles | To replace damaged bulkheads on junction boxes. |

*Table Suggested Cable Spares*

Having the right cable on the shelf turns a 4-hour down situation into a 15-minute repair.

## DECISION TREE REFERENCE

A cable and wiring troubleshooting decision tree is available at BootsOnTheGroundTech.com for download and printing.

**Pro Tip**: The wiggle test is worth more than any cable tester. If the signal drops when you flex the cable, you've found your problem—no fancy equipment required.

**Shop-Floor Wisdom**: When a sensor "keeps failing," look at the cable before you condemn the sensor. Cables fail more often than sensors do.

# CHAPTER 22
# PLC LANGUAGES & LOGIC

Every sensor, every output, every decision in an automated system flows through the PLC program. When you're troubleshooting field devices, you'll inevitably end up staring at PLC logic—trying to figure out why a valve won't open, why a motor won't start, or why an alarm won't clear.

You don't need to be a programmer to troubleshoot effectively. But you do need to understand how PLC logic works, how to read the major programming languages, and how to trace through code to find what's holding up your machine.

This chapter covers the essentials—enough to make you dangerous with a laptop connected to any PLC.

## THE SCAN CYCLE

A PLC program runs in a continuous loop: Read inputs (scan all input cards into memory) → Execute logic (run program top to bottom, evaluate conditions, set outputs) → Write outputs (update all output cards) → Repeat. This cycle happens continuously, typically every 5-50 milliseconds depending on program size and processor speed.

When you go online and watch logic execute, you're seeing a snapshot of what the processor did during the last scan. Fast or transient events can be missed. Changes you make to outputs (forcing bits) take effect on the next scan—not instantly.

## THE IEC 61131-3 LANGUAGES

IEC 61131-3 is the international standard that defines programming languages for PLCs. It standardized what was previously a chaotic mix of proprietary formats—every manufacturer had their own approach. Now, whether you're working on Rockwell, Siemens, Schneider, or Beckhoff, you'll encounter these same five languages. Understanding which language does what helps you read existing code and troubleshoot logic problems faster.

| Language | Abbr. | Type | Primary Use | When You'll See It |
|---|---|---|---|---|
| Ladder Diagram | LD | Graphical | Discrete logic; relay replacement. | North American manufacturing. |
| Function Block | FBD | Graphical | Process control; analog loops. | Process plants; Siemens/ABB. |
| Structured Text | ST | Textual | Complex math; string handling. | Beckhoff; Rockwell math blocks. |
| Sequential Chart | SFC | Graphical | Batch processes; step sequences. | Packaging; CNC machines. |
| Instruction List | IL | Textual | Legacy/Low-level assembly. | Rare in modern systems. |
| Rodents | | Visible chew marks. | Food plants; warehouses. | Use armored cable or conduit. |
| Ambient Light | | Fails in direct sunlight. | Receiver saturation from sun/lights. | Add a sun shield or hood. |

*Table IEC6131*

Each language has strengths for specific tasks. Ladder logic dominates discrete manufacturing because electricians already understand relay logic. Function Block excels at process control and analog loops. Structured Text handles complex math and string manipulation. Sequential Function Chart manages batch processes and state machines. Instruction List appears rarely in modern systems but still exists in legacy code.

The sections below focus on what you need to troubleshoot existing programs, not write new ones.

# LADDER DIAGRAM (LD)

Ladder logic is the dominant language in discrete manufacturing, especially in North America. It represents logic as horizontal "rungs" between two vertical power rails—designed to look like the relay logic diagrams that electricians already knew.

Read ladder logic left to right: Contacts in series = AND logic (all must be true). Contacts in parallel = OR logic (any can be true). The coil at the right end is the result.

| Symbol | Name | Function | When It's True | When You'll See It |
|---|---|---|---|---|
| –[ ]– | NO Contact | Examines a bit for ON. | Bit is 1 (High). | North American manufacturing. |
| –[/]– | NC Contact | Examines a bit for OFF. | Bit is 0 (Low). | Process plants; Siemens/ABB. |
| –( )– | Output Coil | Sets bit ON if rung is true. | Rung is True. | Beckhoff; Rockwell math blocks. |
| –(L)– | Latch (OTL) | Sets bit ON; stays ON. | Rung goes True. | Packaging; CNC machines. |
| –(U)– | Unlatch (OTU) | Sets bit OFF. | Rung goes True. | Rare in modern systems. |
| –[ONS]– | One-Shot | True for one scan only. | False → True transition. | |
| Ambient Light | Fails in direct sunlight. | Receiver saturation from sun/lights. | Add a sun shield or hood. | |

*Table Ladder Logic*

Online monitoring: When connected online, most programming software highlights power flow. Green (or highlighted) = condition is true, power is flowing. Gray (or dim) = condition is false, power is blocked. Follow the highlights to see exactly where the logic is blocked.

**Troubleshooting Ladder Logic**

When outputs don't turn on, trace power flow backward from the coil. If the coil isn't energized, at least one contact in that rung is false. Find which contact is blocking power—it's either a physical input that's not active, a timer that hasn't timed out, or an internal bit controlled by logic elsewhere in the program.

## FUNCTION BLOCK DIAGRAM (FBD)

FBD represents logic as interconnected blocks—inputs flow in on the left, outputs flow out on the right, data flows left to right through the network. Common in process control, motion control, and European-based systems. Values appear on the wires between blocks so you can see exactly what's entering and leaving each block online.

| Block | Function | Output Condition | Meaning |
|---|---|---|---|
| AND | Logical AND | True only if ALL inputs are true. | Calibrated zero |
| OR | Logical OR | True if ANY input is true. | 1/4 Scale |
| NOT | Inversion | Output is opposite of input. | Midpoint check |
| PID | Control Loop | Calculates output based on Setpoint vs. PV. | 3/4 Scale |
| EQ / GT | Comparison | True when values match or exceed. | Full scale |
| ADD / DIV | Math | Arithmetic result of inputs. | |

*Table Function Block Diagram*

**Troubleshooting Structured Text:** look for the variable assignments that control your output; trace backward through the conditions that gate those assignments; watch variable values online in a watch window—most platforms support real-time ST variable monitoring.

When troubleshooting FBD, follow the data flow from inputs through each block to the output. If a block's output is wrong, check its inputs first—garbage in, garbage out. PID blocks are common culprits: verify setpoint and process variable values are in the correct range and units. Math blocks fail when inputs are out of bounds or when divide-by-zero occurs. Most platforms show live values on each wire when online.

## SEQUENTIAL FUNCTION CHART (SFC)

SFC organizes logic into steps and transitions. Each step contains actions; transitions define the conditions required to advance to the next step. Common in batch processes, packaging lines, CNC machines, and any sequential operation with defined phases.

Troubleshooting SFC: identify which step is currently active (usually highlighted or shown in a step register); examine the transition below that step—what condition is it waiting for?; trace the blocking condition to its source.

## TIMERS AND COUNTERS

| Timer | Behavior | Common Use |
|---|---|---|
| TON (On-Delay) | Counts when Enable is True. | Input debouncing; safety delays. |
| TOF (Off-Delay) | Starts when Enable goes False. | Cooling fans; delayed shutdowns. |
| RTO (Retentive) | Retains time when paused. | Maintenance timers; hour meters. |
| Insulation | Megohmmeter test (Conductor to Gnd). | Finds moisture or insulation breakdown. |
| Substitution | Swap with a known good cordset. | Quickly proves the cable is the "liar." |

*Table PLC Timer Types*

**Timer troubleshooting:** Check the enable condition (is it true?); check the preset value (is it reasonable—not accidentally 500,000 ms instead of 500 ms?); check the accumulated value (is it counting?); check the done bit (does it ever go TRUE?).

Most timer problems aren't the timer itself—they're the enable condition that never goes true or a reset condition that fires unexpectedly. Watch the timer online: if the accumulated value isn't incrementing, the enable bit is false. If it resets before timing out, something is triggering the reset condition. Follow the logic backward from there.

**Counter troubleshooting:** Is the count input pulsing? Is the accumulated value changing? Is something unexpectedly resetting the counter? Is the preset correct?

Watch the counter online while the input pulses. If the accumulated value doesn't increment, the count input isn't reaching the counter—check the physical sensor and wiring. If the counter increments but unexpectedly resets, something is triggering the reset condition—find it in the logic. If the counter stops at a value below the preset, check if something is disabling the count enable bit. Counters are simple—the problem is usually in the conditions that control them, not the counter itself.

| Counter | Behavior | Done Bit (DN) |
|---|---|---|
| CTU (Up) | Increments on rising edge. | True when Acc $\ge$ Preset. |
| CTD (Down) | Decrements on rising edge. | True when Acc $\le$ 0. |
| RTO (Retentive) | Retains time when paused. | Maintenance timers; hour meters. |
| Insulation | Megohmmeter test (Conductor to Gnd). | Finds moisture or insulation breakdown. |
| Substitution | Swap with a known good cordset. | Quickly proves the cable is the "liar." |

*Table PLC Counters*

## FORCES: THE DOUBLE-EDGED SWORD

Forces allow you to override the program and manually set an input or output bit to a specific state. Invaluable for testing—dangerous if left in place.

An output force bypasses all programmed interlocks and safety logic. A force left active from a previous troubleshooting session is one of the most common causes of "impossible" machine behavior.

# PLC Languages & Logic

> ⚠️ **BEFORE YOU GO DEEPER: BACK UP THE PROGRAM FIRST**
>
> Before making any program changes, forcing outputs, or modifying timers and presets during troubleshooting, upload and save a copy of the current program to your laptop. Use the same filename with a date and time stamp. This takes two minutes and has saved technicians from overwriting a running production program with an earlier version more times than can be counted. If the machine runs after your troubleshooting session, that backup is your proof of what you found and what you changed.

Check the force table FIRST when troubleshooting unexplained behavior:

| Platform | Where to Find Forces |
|---|---|
| Rockwell (Logix) | Controller Properties → General → Controller Tags → Force Table; check the physical Force LED on the CPU. |
| Siemens TIA | Online & Diagnostics → Force values. |
| Siemens Step 7 | PLC menu → Force Variables. |
| Mitsubishi | Device Monitor → Entries in red or marked with "F". |
| Omron | Watch Window → Forced values indicated. |
| Beckhoff / Codesys | Online → Force Values table. |

*Table PLC Force Tables by Vendor*

Best practices: Remove all forces when done testing. Check the force table at shift change. Never leave forces active overnight without documenting and notifying the oncoming shift.

> **Field note**: A gripper would randomly drop parts. Electricians replaced solenoids, checked valves, rewired the connector. Finally someone checked the force table—an output was forced ON from testing two weeks prior. The force intermittently conflicted with the program's commands. Remove the force, problem solved. Total wasted time: 4 hours. Total wasted parts: $400. The Force Enable LED on the processor front panel is your first check—if it's ON, there are active forces somewhere.

## COMMON TROUBLESHOOTING SCENARIOS

| Scenario | First Check | Second Check | Third Check |
|---|---|---|---|
| Output won't turn ON | Is the rung highlighted (True)? | If False: find the open contact blocking the rung. | If True: check for forces or the same bit being overwritten elsewhere. |
| Output won't turn OFF | Check for forced ON bits. | Search for "Latches" (OTL) missing an "Unlatch" (OTU). | Verify field device isn't mechanically stuck. |
| Sequence Stuck | Identify the active step. | Find the transition condition (what is it waiting for?). | Trace the missing input to the field sensor. |
| Erratic Behavior | Check for old forces left behind. | Trend inputs to find electrical noise. | Check for recently changed code or uninitialized data. |
| Power Cycle Fail | Check for non-retentive bits. | Review the "First Scan" initialization routine. | Look for latch coils that need a manual reset. |

*Table Common PLC Troubleshooting Scenarios*

## THE DIAGNOSTIC BUFFER: WHAT FAULTED, WHEN, AND IN WHAT ORDER

Most modern PLCs maintain an onboard diagnostic buffer—a time-stamped log of faults, mode changes, and system events. For intermittent or recurring problems, this is your first stop before looking at a single rung of logic.

**Rockwell (Studio 5000 / RSLogix 5000):** In the Controller Properties window, select the Diagnostics tab. The Controller Log and the Major/Minor Fault History show fault codes with timestamps. Note the fault code number—Rockwell's online help and the L5X export decode them precisely.

**Siemens (TIA Portal):** In the Project tree, navigate to the CPU → Diagnostics → Diagnostic buffer. Events are listed in reverse chronological order with timestamps. The buffer holds the last 100+ events depending on CPU model.

**What to look for:** a fault that clears and resets repeatedly at regular intervals points to a programmatic cause (timer, sequence). A fault that correlates with a specific shift or time of day points to an environmental or load change. A fault that appeared once and stayed points to hardware. The timestamp sequence tells you what happened first—which is almost always more useful than the fault code itself.

**Pro Tip:** Before your first shift on an unfamiliar machine, read the diagnostic buffer. You'll see every fault it has thrown in recent history. That history is a map of the machine's chronic problems.

## FINDING YOUR WAY IN UNFAMILIAR CODE

1. Search before scrolling. Use the search function to find tags, addresses, and comments. Don't scroll through thousands of rungs.
2. Use cross-reference. Select any tag and find every place it's used—reads, writes, all of it. This is the fastest way to understand how a signal flows through the program.
3. Look for comments. Good programmers comment their code. Rung comments, tag descriptions, and program notes are your roadmap.

4. Follow the sequence. Find the step counter or sequence state variable. Watch it as the machine runs. When it stops advancing, find what transition it's waiting for.
5. Find the mode logic. Most machines have modes: Auto, Manual, Setup, Jog. The mode determines which logic is active. Confirm you're in the right mode before troubleshooting.

---

## VENDOR-SPECIFIC NOTES

Each PLC manufacturer uses different terminology and tools for the same troubleshooting tasks. What Rockwell calls "Ctrl+E" for cross-reference, Siemens calls "Cross-Reference" and Beckhoff calls "Find all references." The table below translates common troubleshooting functions across platforms. When you're working on an unfamiliar system, this quick reference tells you where to find the tools you need.

| Platform | Tag System | Cross-Reference | Monitor / Trace | Force Table |
|---|---|---|---|---|
| Rockwell | Named tags. | Ctrl+E | Watch Window / Trends | Controller Properties |
| Siemens TIA | Symbolic. | Edit → Cross-Ref | Watch Table / Trace | Online & Diagnostics |
| Siemens S7 | Fixed (I/Q/M). | Cross-Reference | Variable Table (VAT) | Force Variables |
| Mitsubishi | Fixed (X/Y/M/D). | Device Usage List | Device Monitor | Device Monitor |
| Beckhoff | ST Variables. | Find all references | Inline Monitoring | Force Values table |

*Table PLC Vendor Specific Notes*

## REAL-WORLD EXAMPLES

### Example 1: The Missing Permissive

**Problem**: Conveyor won't start in Auto mode. Works fine in Manual.

**Finding**: In Auto mode, one contact on the conveyor output rung was false: "Upstream_Ready." Upstream_Ready is set by the upstream machine's PLC via network. Upstream PLC was in fault state and not sending the Ready signal.

**Fix**: Cleared fault on upstream PLC. Conveyor started.

## Example 2: The Hidden Force

**Problem**: Safety light curtain muted—but mute doesn't release when operator clears the field. All conditions for un-muting are satisfied in logic, but output stays ON.

**Finding**: Output was forced ON. A technician had forced it during testing and forgot to remove the force.

**Fix**: Removed the force. Mute operates correctly.

## Example 3: The Timer That Wasn't

**Problem**: Part presence sensor works, but filling doesn't start. Machine waits for "part stable" timer that never completes.

**Finding**: Timer preset had been accidentally changed from 500 ms to 500,000 ms (500 seconds) during a recent program edit.

**Fix**: Corrected preset to 500 ms.

## Example 4: The Last One Wins

**Problem**: Indicator light comes on for a split second then goes off, even though the alarm condition is active.

**Finding**: A second rung later in the program was turning the indicator OFF unconditionally—programmer had added it for testing and forgot to remove it. Since it was later in the scan, it overwrote the first rung's result.

**Fix**: Removed the erroneous rung.

## DECISION TREE REFERENCE

A PLC troubleshooting decision tree is available at BootsOnTheGround-Tech.com for download and printing.

💡 **Pro Tip**: Before you blame the program, check the forces. Before you blame the field device, verify the PLC is actually seeing what you expect. Troubleshoot systematically, not randomly.

⚡ **Shop-Floor Wisdom**: The logic is almost always right. What's usually wrong is the assumptions—wrong mode, missing permissive, force left active, or a field device not doing what you think it's doing.

# PART IV
# PUTTING IT TOGETHER

# CHAPTER 23
# CASE FILES

Theory is useful. Principles are important. But nothing teaches like real problems on real machines with real pressure to get them running.

These case files are drawn from thirty years of field experience—automotive plants, chemical facilities, food processing, packaging lines, and everything in between. Names and some details have been changed, but the problems are real. The lessons are real. And the mistakes are definitely real.

Each case follows the SITVD framework: Symptom, Isolate, Test, Verify, Document. Use these stories to build your pattern recognition. The next time you see something similar, you'll have a head start.

## CASE FILE #1: RTD THERMAL DRIFT

**The Setup:** Food processing kettle; 185°F safety hold (HACCP). Pt100 RTD to 4-20 mA transmitter.

**[S] Symptom:**

HMI indicated a static 142°F while the steam valve remained at 100% output. Process temperature failed to rise on the HMI for 20 minutes despite active heating.

**[I] Isolate:**

- **Mechanical:** Steam pressure (45 PSIG) and condensate return verified operational.
- **Instrumentation:** Transmitter display matched the HMI (142°F), isolating the issue to sensor scaling rather than the PLC or network.

**[T] Test:**

- **Reference Check**: A calibrated digital probe measured the actual product at 187°F.
- **Bench Test**: The RTD measured at 156 Ω. At 187°F, a Pt100 should measure 133.2 Ω.

**[V] Verify:**

Replaced the RTD element. The new sensor matched the reference probe within 1°F. The steam valve maintained setpoint.

**[D] Document:**

- **Root Cause:** RTD element drifted due to chronic thermal shock.
- **Corrective Action:** Replaced sensor

## CASE FILE #2: MAG METER AIR ENTRAINMENT

**The Setup:** Chemical facility; caustic soda batching via magnetic flow meter. Accuracy is critical for reaction completion and recipe integrity.

**[S] Symptom:** Flow readings were erratic, oscillating rapidly between 0 and 120 GPM despite a steady pump speed. Downstream batch results were inconsistent.

### [I] Isolate:

- **Mechanical:** A downstream sample valve was opened to confirm physical flow. A steady stream verified the pump and piping were not surging.
- **Instrumentation:** The local transmitter display matched the HMI fluctuations. Diagnostic menus revealed an intermittent "Empty Pipe" alarm.

### [T] Test:

- **Physical Audit:** Inspected the piping geometry. The mag meter was installed at the hydraulic high point of the run, with the discharge dropping vertically into a mixing tank.
- **Process Observation:** The intermittent alarm coincided with air slugs accumulating at the high point due to minor pump cavitation.

**[V] Verify:** An air elimination valve was temporarily installed upstream of the meter. Once the accumulated air was bled, flow readings stabilized at a consistent 45 GPM.

**[D] Document:**

- **Root Cause:** Improper installation at a piping high point, leading to air entrapment and "Empty Pipe" false-positives.
- **Risk Exposure:** Batch contamination and chemical waste.
- **Corrective Action:** Relocated mag meter to a lower elevation in the piping run during the next scheduled shutdown.

———

## CASE FILE #3: THE PROXIMITY SENSOR FLEX FATIGUE

**The Setup:** Automotive stamping; weld fixture part-seat detection. Inductive proximity sensor provides a safety interlock for robotic weld cycles.

**[S] Symptom:** Intermittent "Part in Position" faults. Robot cycle halts awaiting input despite part presence. Fault frequency: ~10 occurrences per shift.

**[I] Isolate:**

- **HMI:** Confirmed PLC awaiting input.
- **Sensor Local LED:** Observed in OFF state with part present.
- **Physical Alignment:** Sensor face clean; steel target within the 8mm rated range (measured gap: 3mm).

**[T] Test:**

- **Cable Stress Test:** Flexed the integral sensor cable near the housing head while the part was present.
- **Observation:** The local LED flickered during cable manipulation, indicating internal conductor failure.

**[V] Verify:** Installed new proximity sensor. Cycled the fixture 50 times to confirm signal continuity through the full range of motion.

**[D] Document:**

- **Root Cause:** Internal conductor fatigue caused by repeated bending at the sensor head.
- **Risk Exposure:** Unscheduled downtime and robot cycle interruptions.
- **Corrective Action:** Replaced sensor; installed cable carrier and expanded service loop to manage bending radius.

## CASE FILE #4: ANALOG LOOP SIGNAL LEAKAGE

**The Setup:** Water treatment facility; filter backwash pressure monitoring. 4-20 mA loop from a 0-50 PSIG pressure transmitter to a PLC input card.

**[S] Symptom:** Automatic backwash cycle failed to terminate. HMI indicated 0 PSIG throughout the cycle. Local transmitter display indicated 22 PSIG, creating a discrepancy between field and control room data.

**[I] Isolate:**

- **Source Check:** Measured loop current at the transmitter terminals. Output was 11.2 mA (~22.5 PSIG), confirming the transmitter was functioning correctly.
- **Destination Check:** Measured loop current at the PLC input card. Input was 4.0 mA (0 PSIG).
- **Isolation:** 7.2 mA of the signal was being diverted between the transmitter and the PLC panel.

**[T] Test:**

- **Sectional Measurement:** Measured current at the intermediate junction box (11.2 mA) and the PLC panel terminal strip (4.0 mA). The loss was isolated to the 200-foot cable run.
- **Physical Audit:** Inspected the cable tray route through the boiler room. A structural tray failure had allowed the instrument cable to rest directly on a high-pressure steam pipe.

**[V] Verify:** Installed new shielded twisted-pair cable. Verified current at the PLC input card matched the transmitter output at 11.2 mA. Automatic backwash control was restored.

**[D] Document:**

- **Root Cause:** Signal leakage due to insulation failure. Thermal damage from a steam pipe created a high-resistance parallel path to ground, diverting loop current.
- **Risk Exposure:** Inefficient water treatment cycles and manual override errors.
- **Corrective Action:** Replaced 200 feet of instrument cable; repaired cable tray and rerouted away from high-heat sources.

———

# CASE FILE #5: THE SCALE THAT COULDN'T ZERO

**The Setup:** High-speed packaging line; in-motion checkweigher. System verifies carton weight at a rate of 60 ppm, with an automatic tare cycle between units.

**[S] Symptom:** The scale failed to zero during the tare interval. With the platform empty, the display indicated a static +1.2 lb offset. High reject rates occurred as compliant cartons were flagged as overweight.

**[I] Isolate:**

- **Surface Inspection:** Platform surface was verified clean and free of debris.
- **Structural Clearance:** The gap between the weighing platform and the stationary frame appeared clear upon visual inspection.
- **Mechanical Test:** A lateral "push test" was performed on the platform. The weight reading shifted inconsistently (+1.2 lb to +0.8 lb), confirming a mechanical binding rather than an electronic zero-drift.

**[T] Test:**

- **Borescope/Internal Audit:** Inspected the underside of the weighing structure.
- **Observation:** A discarded nylon cable tie (zip tie) was found wedged across the clearance gap between the load-cell-mounted platform and the stationary machine frame.

**[V] Verify:** Removed the cable tie. The scale zeroed immediately. Accuracy was verified across the full span using certified test weights.

**[D] Document:**

- **Root Cause:** Mechanical bridging. A foreign object transmitted force between the weighed and unweighed structures, creating a false load.
- **Risk Exposure:** Significant product waste (false rejects) and potential calibration integrity issues.
- **Corrective Action:** Removed obstruction; inspected overhead cable management to prevent further debris falls.

———

# CASE FILE #6: VALVE ACTUATOR SPRING FAILURE

**The Setup:** Beverage bottling line; syrup mixing tank. A pneumatic flow-control valve utilizes 24VDC solenoid control and mechanical limit switch feedback.

**[S] Symptom:** Syrup overflow occurred despite a "CLOSED" command from the PLC. The HMI indicated the valve was closed and verified via feedback, but physical flow continued until an upstream manual isolation valve was closed.

**[I] Isolate:**

- **Feedback Discrepancy:** The physical actuator position indicator was observed in the "OPEN" state. However, the PLC input for the "Closed" limit switch remained TRUE.
- **Point-to-Point Check:** Measured the limit switch state at the junction box. The switch contacts were correctly open (actuator not at limit), but 24VDC was present on the wire pair returning to the PLC.
- **Source of Error:** A physical jumper wire was discovered across the terminal strip, bypassing the field device.

**[T] Test:**

- **Mechanical Audit:** With the solenoid de-energized and air exhausted, the valve failed to return to its fail-safe position.
- **Observation:** Disassembly of the actuator housing revealed a fractured internal return spring, preventing the mechanical stroke to the closed position.

**[V] Verify:** Removed the unauthorized jumper. Installed a new actuator spring. Cycled the valve five times to verify that the PLC "Closed" feedback bit accurately tracked with the mechanical position of the actuator.

**[D] Document:**

- **Root Cause:** Double-fault condition. A mechanical spring failure was masked by an undocumented electrical jumper left from a previous shift's temporary bypass.
- **Risk Exposure:** Product loss due to tank overflow and safety risk from unmonitored material flow.
- **Corrective Action:** Replaced actuator spring; removed signal bypass; updated the "Temporary Modification" log.

# CASE FILE #7: SIGNAL CROSSTALK

**The Setup:** Distribution center; automated barcode scanning station. A retroreflective photoeye triggers the scanner as cartons interrupt the beam between the sensor and the reflector.

**[S] Symptom:** Inconsistent scanner triggering. Approximately 30% of cartons passed the sensor without activating the trigger signal. No correlation was found regarding carton size or weight.

**[I] Isolate:**

- **Visual Audit:** Observed the sensor status LED during production. Matte cardboard cartons triggered the sensor reliably. Glossy, shrink-wrapped cartons passed without the sensor registering a beam break.
- **Observation:** The glossy surface of certain cartons appeared to be reflecting the emitted light back to the sensor receiver.

**[T] Test:**

- **Surface Manipulation:** A piece of matte cardboard was applied to the surface of a shrink-wrapped carton. The sensor triggered correctly. Upon removal of the cardboard, the same carton failed to trigger.
- **Diagnosis:** The non-polarized sensor was unable to distinguish between the light returning from the dedicated reflector and the specular reflection from the glossy product.

**[V] Verify:** Installed a polarized retroreflective sensor and a matching polarized reflector. The miss rate dropped to 0%. The sensor now successfully ignored the unpolarized reflection from the glossy shrink-wrap.

**[D] Document:**

• **Root Cause:** Optical crosstalk. A standard retroreflective sensor detected the specular reflection from the product surface as a "valid" return signal, masking the beam break.

• **Risk Exposure:** Sorting errors, increased manual rework, and degraded system throughput.

• **Corrective Action:** Upgraded to polarized optical components for all glossy product lines.

---

# CASE FILE #8: EMI AND SHIELD GROUND LOOPS

**The Setup:** Paper mill; basis weight measurement via nuclear beta gauge. A 4-20 mA loop controls a critical stock flow valve for paper density regulation.

**[S] Symptom:** Severe signal oscillation (±15%) at the PLC input. The instability made automatic process control impossible, forcing operators into manual mode and resulting in inconsistent product quality.

**[I] Isolate:**

- **Source Check:** The gauge local display was verified as stable (55.2 gsm ± 0.1).
- **Destination Check:** PLC raw counts showed high-frequency fluctuations. The signal was clean at the transmitter but corrupted at the controller.
- **Environmental Change:** Maintenance confirmed a recent Variable Frequency Drive (VFD) replacement on the main machine drive.

**[T] Test:**

- **Shield Audit:** Inspected the instrument cable shielding. The drain wire was found terminated to ground at both the transmitter housing and the PLC cabinet.
- **Diagnosis:** Dual-point grounding created a ground loop. The potential difference between the two ground points allowed current to flow through the shield, effectively turning the cable into an antenna for the high-frequency switching noise of the new VFD.

[V] **Verify:** Disconnected ("lifted") the shield ground at the field transmitter, leaving it grounded only at the PLC common ground point. Signal noise was reduced by 90%. A filtering capacitor was installed at the PLC input card to mitigate the remaining residual EMI.

### [D] Document:

- **Root Cause:** Shield ground loop. The dual-ground configuration induced electromagnetic interference into the signal conductors, a condition exacerbated by the high switching frequency of the newly installed VFD.
- **Risk Exposure:** Off-spec product and loss of automated process control.
- **Corrective Action:** Isolated signal shield to a single-point ground; scheduled rerouting of instrument cabling away from 480V power feeders.

---

# WHAT THESE CASES HAVE IN COMMON

Reviewing these eight case studies reveals consistent patterns in industrial failure modes. Understanding these themes is the first step toward developing professional intuition.

### 1. The Fallacy of the Obvious

Primary symptoms are frequently decoys. Effective troubleshooting requires looking past the immediate failure to the underlying cause:

- **Sensor Drift (Case 1):** Masked as a utility/heating failure.
- **Hydraulic Geometry (Case 2):** Masked as a flow meter failure.
- **Mechanical Bridging (Case 5):** Masked as a calibration/load cell error.
- **Signal Bypassing (Case 6):** Masked as a valve actuator failure.

### 2. The Power of Independent Verification

Data on an HMI is a representation of reality, not reality itself. Every successful diagnosis involved a secondary standard:

- **Reference Standards:** Using a calibrated probe to disprove in-situ sensor data.
- **Physical Validation:** Opening a sample valve to confirm hydraulic presence.
- **Mechanical Stress Tests:** Using a "push test" or "wiggle test" to reveal binding or fatigue.
- **Visual Audit:** Confirming that physical position contradicts electronic feedback.

### 3. Impact of Environmental Factors

Instrumentation does not exist in a vacuum. The physical surroundings are often the root cause:

- **Kinematics:** Mechanical motion leading to cable flex fatigue (Case 3).
- **Thermal Stress:** Proximity to steam lines or chronic thermal shock (Cases 4 & 1).
- **Material Properties:** specularity and reflectivity affecting optical paths (Case 7).
- **Electromagnetic Interference:** High-frequency noise from VFD switching (Case 8).

### 4. Temporal Context

Most "random" failures have a chronological trigger. Always ask, "What changed, and when?"

- **Duty Cycles:** Cumulative fatigue over hundreds of thousands of operations.
- **Legacy Modifications:** Undocumented jumpers left from previous shifts.
- **System Upgrades:** New equipment (like VFDs) altering the baseline environment.

### 5. Validation of the SITVD Framework

The **SITVD** methodology provides a repeatable path to resolution:

- **[S] Symptom:** Precisely define the discrepancy.
- **[I] Isolate:** Segment the loop to narrow the search area.
- **[T] Test:** Use measurements to confirm or deny each hypothesis.
- **[V] Verify:** Ensure the system returns to its design state.
- **[D] Document:** Close the loop to prevent future recurrence.

## DEVELOPING A PERSONAL KNOWLEDGE BASE

Every troubleshooting engagement is an opportunity to expand your internal library of case files. When a resolution is reached, formally documenting the event serves two purposes: it solidifies the logic in your own memory and provides a searchable record for future failures.

### The Post-Action Review:

- **Symptom Definition:** What was the specific data/physical discrepancy?
- **Isolation Path:** Which components were eliminated from suspicion?
- **The "Eureka" Moment:** What was the definitive test that revealed the cause?
- **Resolution:** What specific action restored the design state?
- **Process Improvement:** How can the SITVD approach be optimized for this failure mode in the future?

Over time, these records evolve into your most valuable professional asset. True expertise is simply the ability to perform rapid pattern-matching against a deep archive of prior experiences. That process begins with rigorous observation and disciplined note-taking.

---

> 💡 **Pro Tip:** The solution to most "impossible" problems is something simple that nobody checked. Foreign objects, forgotten jumpers, wrong sensor type, misconfigured parameters—the exotic failures are rare. Check the basics first.
>
> ⚡ **Shop-Floor Wisdom:** If it worked yesterday and doesn't work today, something changed. Find out what changed, and you'll probably find your problem.

---

# CLOSING THOUGHTS

You have reached the end of this volume. Whether you read it cover-to-cover or utilized specific chapters to solve immediate field problems, you now possess a framework for troubleshooting that many technicians take a decade to develop. The **SITVD** method, the diagnostic chain, and the isolation principle are not academic theories—they are the distilled reality of three decades spent maintaining critical infrastructure when conventional solutions failed.

## THE INEVITABILITY OF ERROR

I opened this book with the statement that you are going to screw up, and I close with it here. Understanding this is liberating. Every expert has a mental archive of embarrassing oversights—hours spent chasing a "faulty" PLC card that was actually a blown fuse, or replacing a sensor three times before noticing a simple wiring error.

The distinction between a novice and a professional is not the absence of mistakes; it is the speed at which the professional recognizes the error, corrects it, and updates their checklist to ensure it never happens again.

## TRUST YOUR PROCESS

Under the pressure of downtime, the temptation to skip steps is immense. You will feel the urge to swap parts based on a "hunch" because doing something feels better than methodically measuring. **Resist that temptation.**

The SITVD framework is built for high-pressure environments. When the situation becomes chaotic, let the script lead you:

- **[S] Symptom:** Define the specific discrepancy.
- **[I] Isolate:** Segment the loop and cut the problem in half.
- **[T] Test:** Verify every assumption with a calibrated measurement.
- **[V] Verify:** Confirm the resolution through full-cycle operation.
- **[D] Document:** Record the finding for your future self and your team.

## THE METER DOESN'T LIE

Your multimeter is the ultimate arbiter of truth. It does not care about assumptions, opinions, or what "should" be happening. It reports reality.

- When a technician says the wiring is "fine," the meter verifies continuity.
- When the HMI claims an input is active, the meter verifies the voltage at the terminal.
- When a transmitter reports 50%, the meter confirms if 12.0 mA is actually flowing.

Trust your meter more than the HMI, the operator, or your own assumptions. However, remember that the meter only tells the truth if it is configured correctly for the task.

## UNDERSTAND THE ENTIRE CHAIN

Every signal flows through a continuous chain: the physical process, the field device, field wiring, the I/O card, PLC logic, and the HMI. A break anywhere in this chain produces a symptom at the end of it. To be an elite troubleshooter, you must see the chain, not just the components. Do not waste time on sensor calibration when the problem is a scaling error in the PLC.

## FIELD EXPERIENCE IS IRREPLACEABLE

This book provides the framework, but nothing replaces time in front of a live machine. Every failure mode you solve adds to your pattern-matching database. The best troubleshooters are those who have seen the most problems and—crucially—paid the closest attention.

## THE CRAFT AND THE COMMUNITY

Industrial troubleshooting is a craft that blends technical science with professional judgment. Respect the work. Take pride in the "tough wins."

Most importantly, **teach what you learn.** The best way to cement your own knowledge is to mentor someone else. The technician who hoards knowledge is a bottleneck; the technician who shares knowledge builds a resilient team.

---

## WHAT'S NEXT

This book is the first in the Boots on the Ground series. It covers instrumentation and signals—the foundation of industrial troubleshooting. But there's more ground to cover:

- **Book 2: PLCs & Controls**—Deeper into ladder logic, function blocks, structured text, and control system architecture
- **Book 3: Communications & Networking**—Ethernet/IP, Profinet, Modbus, serial communications, and network troubleshooting

Each book builds on the others, but each stands alone. Take what you need, when you need it.

## USE THIS BOOK

This is a field manual, not a trophy for your shelf. Dog-ear the pages. Highlight the tables. Photocopy the SITVD checklists and laminate them for your toolbox. When you are standing in front of a machine at 2:00 AM, I want this book to be the resource that gets the line running.

Thank you for your time, your dedication to the craft, and for being the person who stays until the problem is solved.

**Now, go fix something.**

**Bob**

*Thirty years in the field and still learning*

**www.BootsOnTheGroundTech.com**

**Code:** COLDSPRAY (For additional resources and decision tree downloads)

Please leave a review wherever you purchased the book, if you found this book to be useful.

# APPENDICES

APPENDIX A: Pocket Math for the Field

APPENDIX B: Discrete Signal Troubleshooting

APPENDIX C: Analog Loop (4-20 mA) Troubleshooting

APPENDIX D: Inductive & Capacitive Proximity Sensors

APPENDIX E: Photoeye & Optical Sensor Troubleshooting

APPENDIX F: Cable Integrity and Wiring Standards

APPENDIX G: Temperature Sensor (RTD/TC) Troubleshooting

APPENDIX H: The Essential Field Technician's Toolkit

APPENDIX I: ISA Instrument Tagging & Symbol Reference

APPENDIX J: Common PLC Fault & Error Code Reference

# APPENDIX A: POCKET MATH FOR THE FIELD

You don't need to be an engineer to do field math. But you do need a handful of formulas and conversions that come up over and over again. This appendix is a quick reference—the math you'll actually use, stripped down to what matters.

Keep this section bookmarked. Or photocopy it and keep them in your backpack.

**A Note on the Layout:** You will find blank pages throughout this appendix. This isn't a printing error; I've left that space for you to use. Every plant has its own quirks—specific scaling constants for unique transmitters, custom PLC overflow values, or specific motor nameplate data. Use those blank pages to log your plant-specific math. This book is a tool; make it yours.

**OHMS LAW | POWER CALCULATIONS | AC RMS**

## 1. DC FUNDAMENTALS (OHM'S LAW)

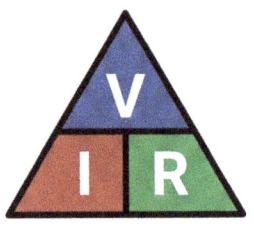

$$V = I \times R, \quad I = V/R, \quad R = V/I$$

## 2. THE POWER RELATIONSHIP

$$P = V \times I$$
$$P = I^2 \times R$$
$$P = V^2/R$$

These three standard Power formulas are algebraically derived directly from Ohm's Law (**V=IR**). This section visualizes the direct mathematical flow from basic DC fundamentals into mechanical work (Watts).

## 3. AC RMS (SINE WAVE REFERENCE)

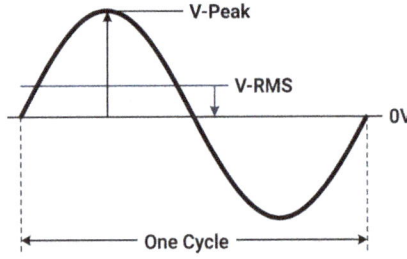

**FIELD NOTE:** All AC power calculations MUST use the RMS (Root Mean Square) value.

### 4-20 mA LOOP CALCULATIONS

Analog signals are linear. If you understand one scaling problem, you understand them all. The numbers change. The math doesn't.

## Quick Reference: The "Rule of Fours"

*Memorize these five points. If your reading is between them, you can estimate the process value instantly.*

| mA Reading | Percent (%) | Signal State |
|---|---|---|
| 20.0 mA | 100% | Upper Range Value (URV) |
| 16.0 mA | 75% | 3/4 Scale |
| 12.0 mA | 50% | Mid-scale |
| 8.0 mA | 25% | 1/4 Scale |
| 4.0 mA | 0% | Lower Range Value (LRV) |
| 3.8 mA | -1.25% | Typical Low-Fail / NAMUR NE43 |
| < 3.5 mA | FAULT | Open Loop / Dead Transmitter |

## INSTRUMENTATION & SIGNALS

### THE MASTER FORMULAS

$$\% \quad \left( \frac{mA - 4}{16} \right) \times 100$$

$$mA \quad \left( \frac{\%}{100} \times 16 \right) + 4$$

### 1. mA to Percent Conversion

$$\% = \left( \frac{mA - 4}{16} \right) \times 100$$

**Example (at 14 mA):**

$$\% = \left( \frac{14 - 4}{16} \right) \times 100 = \mathbf{62.5\%}$$

### 2. Percent to mA Conversion

$$mA = \left( \frac{\%}{100} \times 16 \right) + 4$$

**Example (at 50%):**

$$mA = \left( \frac{50}{100} \times 16 \right) + 4 = \mathbf{12.0 \text{ mA}}$$

# SCALING TO ENGINEERING UNITS

**Master Scaling Formula**

$$Value = \left(\frac{Sig_{In} - Sig_{Min}}{Sig_{Max} - Sig_{Min}}\right) \times (EU_{Max} - EU_{Min}) + EU_{Min}$$

**Example 1: Solving for Pressure (PSI)**

Given: 0–100 PSI transmitter reading 14 mA.

$$Value = \left(\frac{14 - 4}{20 - 4}\right) \times (100 - 0) + 0$$

$$Value = (0.625) \times 100 = \mathbf{62.5 \text{ PSI}}$$

**Example 2: Solving for Required Current (mA)**

Given: What mA signal should a 0–500°F probe send at 300°F?

$$mA = \left(\frac{Value_{In} - EU_{Min}}{EU_{Max} - EU_{Min}}\right) \times (Sig_{Max} - Sig_{Min}) + Sig_{Min}$$

$$mA = \left(\frac{300 - 0}{500 - 0}\right) \times (16) + 4$$

$$mA = (0.6 \times 16) + 4 = \mathbf{13.6 \text{ mA}}$$

**Example 3: Checking for "Over-Range" (High Fail)**

Given: A transmitter is pegged at 20.8 mA. What is the indicated value?

$$Value = \left(\frac{20.8 - 4}{16}\right) \times (100 - 0) + 0 = \mathbf{105 \text{ PSI}}$$

Note: 20.8 mA is the standard high-limit standard for most instrumentation.

**Example 4: 1/4 Scale Verification**

Given: A 0–200 GPM flow meter is reading 8.0 mA. What is the flow?

$$Value = \left(\frac{8 - 4}{16}\right) \times (200 - 0) + 0 = \mathbf{50 \text{ GPM}}$$

Rule of Thumb: 8 mA is always 25% of the engineering span.

# 4–20 mA LOOP RESISTANCE

## LOOP RESISTANCE (DRIVE LIMIT)

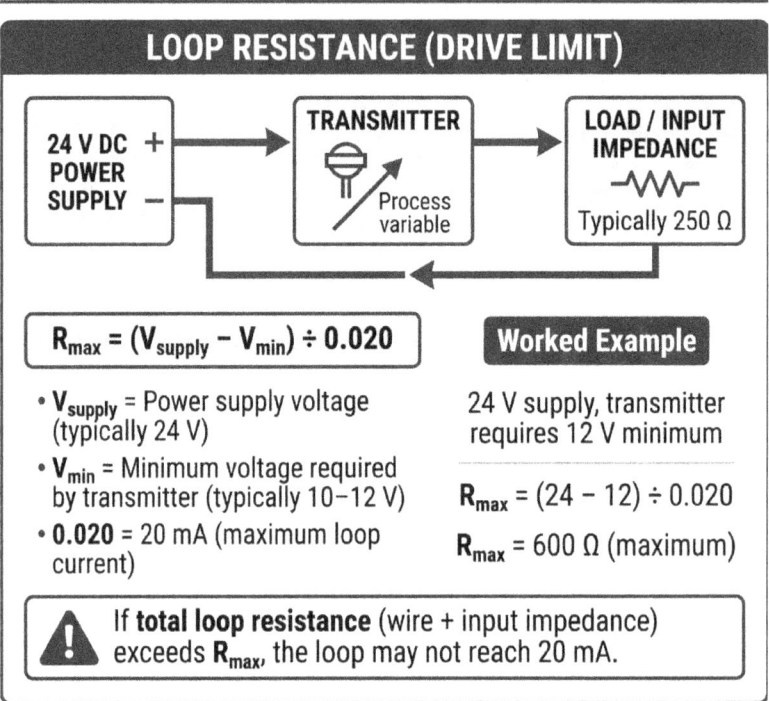

$$R_{max} = (V_{supply} - V_{min}) \div 0.020$$

**Worked Example**

- $V_{supply}$ = Power supply voltage (typically 24 V)
- $V_{min}$ = Minimum voltage required by transmitter (typically 10–12 V)
- 0.020 = 20 mA (maximum loop current)

24 V supply, transmitter requires 12 V minimum

$$R_{max} = (24 - 12) \div 0.020$$

$$R_{max} = 600 \; \Omega \text{ (maximum)}$$

⚠ If **total loop resistance** (wire + input impedance) exceeds $R_{max}$, the loop may not reach 20 mA.

## WIRE RESISTANCE (COPPER)

| Approximate Wire Resistance — Copper | |
|---|---|
| **AWG** | **Ω per 1000 ft** (one conductor) |
| 14 | 2.5 |
| 16 | 4.0 |
| 18 | 6.4 |
| 20 | 10.1 |
| 22 | 16.1 |
| 24 | 25.7 |

**Round-Trip Reminder**

Round trip = 2× cable length (out + back)

**Example Callout**

| 500 ft run of 18 AWG wire |
|---|
| Round trip = 1000 ft |
| Wire resistance ≈ 6.4 Ω |

Wire resistance values are approximate at ~20°C. Resistance increases with temperature.

**Total loop resistance = wire resistance + input impedance**

# APPENDIX A: Pocket Math for the Field

## PRESSURE CONVERSIONS – FIELD REFERENCE

### Common Pressure Unit Conversions

| From | To | Multiply By |
|---|---|---|
| PSI | kPa | 6.895 |
| PSI | bar | 0.0689 |
| PSI | inches $H_2O$ | 27.68 |
| PSI | inches Hg | 2.036 |
| bar | PSI | 14.50 |
| kPa | PSI | 0.145 |
| inches $H_2O$ | PSI | 0.0361 |
| inches Hg | PSI | 0.491 |

### Hydrostatic Pressure (Liquids)

**Pressure (PSI) = 0.433 × SG × Height (ft)**

**Height (ft) = Pressure (PSI) ÷ (0.433 × SG)**

**0.433** = PSI per foot of water
**SG** = Specific gravity (water = 1.0)

**Water (SG = 1.0):**
10 ft = 4.33 PSI

**Caustic (SG = 1.2):**
10 ft = 5.20 PSI

**Liquids vs Gases**
- Hydrostatic formulas apply to liquids
- Gases are compressible
- Pressure does not directly indicate level for gases

- Atmospheric pressure ≈ 14.7 PSIA ≈ 1 bar ≈ 101.3 kPa
- 1 foot of water ≈ 0.433 PSI
- 1 PSI ≈ 2.31 feet of water

APPENDIX A: *Pocket Math for the Field*   233

# TEMPERATURE CONVERSIONS

## CONVERSION FORMULAS

- $°C = (°F - 32) \times \dfrac{5}{9}$

- $°F = \left(°C \times \dfrac{9}{5}\right) + 32$

### QUICK REFERENCE

| °F | °C |
|---|---|
| −40 → | −40 |
| 0 → | −18 |
| 32 → | 0 |
| 68 → | 20 |
| 77 → | 25 |
| 100 → | 38 |
| 212 → | 100 |
| 250 → | 121 |
| 300 → | 149 |
| 350 → | 177 |
| 400 → | 204 |
| 500 → | 260 |

## RTD RESISTANCE (Pt100)

| °C | °F | RESISTANCE (Ω) |
|---|---|---|
| −50 | −58 | 80.3 |
| 0 | 32 | 100.0 |
| 25 | 77 | 109.7 |
| 50 | 122 | 119.4 |
| 100 | 212 | 138.5 |
| 150 | 302 | 157.3 |
| 200 | 392 | 175.8 |

Values are approximate for IEC 60751 Pt100.

### QUICK RTD RULE OF THUMB

- Resistance increases ~0.385 Ω per °C
- At room temperature (25°C), expect ~110 Ω

*APPENDIX A: Pocket Math for the Field* 235

## FLOW CALCULATIONS

### Bucket Test (Volumetric Flow)

**GPM = Volume** (gallons) ÷ **Time** (minutes)

**Example:** 5 gallons collected in 30 seconds
- Time = 30 s = 0.5 min
- GPM = 5 ÷ 0.5 = **10 GPM**

| Common Conversions | | |
|---|---|---|
| From | To | Multiply By |
| GPM → | liters/min → | 3.785 |
| GPM → | m³/hour → | 0.227 |
| liters/min → | GPM → | 0.264 |
| ft³/min (CFM) → | m³/hour → | 1.699 |

### Velocity to Flow (Circular Pipe)

$$Q = A \times V$$

Q = Flow rate (ft³/s or m³/s)
A = Cross-sectional area (ft² or m²)
V = Velocity (ft/s or m/s)

**Pipe Area**

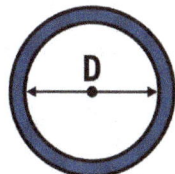

$$A = \pi \times \left(\frac{D}{2}\right)^2$$

$$A = \pi \times D^2 \div 4$$

**Example:** 4-inch pipe, velocity 5 ft/s
- D = 4 in = 0.333 ft
- A = π × (0.333)² ÷ 4 = 0.087 ft²
- Q = 0.087 × 5 = 0.435 ft³/s
- Conversion result shown: ≈ **196 GPM**

(1 ft³/s ≈ 448.8 GPM)

# LEVEL (HYDROSTATIC)

## PRESSURE TO LEVEL (LIQUIDS)

### Level (ft) = Pressure (PSI) ÷ (0.433 × SG)

- 0.433 = PSI per foot of water
- SG = Specific gravity (water = 1.0)

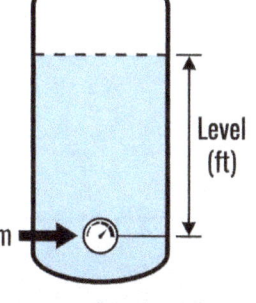

Pressure at bottom → 

Level (ft)

| EXAMPLE 1 — WATER: | EXAMPLE 2 — CAUSTIC: |
| --- | --- |
| 8.66 PSI at bottom of tank | 8.66 PSI, SG = 1.2 |
| Level = 8.66 ÷ (0.433 × 1.0) = 20 ft | Level = 8.66 ÷ (0.433 × 1.2) = 16.7 ft |

**SAME PRESSURE ≠ SAME LEVEL WHEN SG CHANGES.**

## COMMON SPECIFIC GRAVITIES (LIQUIDS)

| Liquid | SG |
| --- | --- |
| • Water | 1.00 |
| • Diesel fuel | 0.85 |
| • Gasoline | 0.74 |
| • Milk | 1.03 |
| • Sulfuric acid (93%) | 1.83 |
| • Caustic soda (50%) | 1.52 |
| • Corn syrup | 1.38 |

Assumptions: Liquid, uniform density, gauge pressure (PSIG).

# SCALING (PLC RAW COUNTS)

## Linear Scaling Formula

$$EU = \frac{(Raw - Raw\_Min) \div (Raw\_Max - Raw\_Min)}{} \times (EU\_Max - EU\_Min) + EU\_Min$$

- **Raw** = current input value (counts)
- **Raw_Min / Raw_Max** = input range endpoints (counts)
- **EU_Min / EU_Max** = engineering unit range endpoints

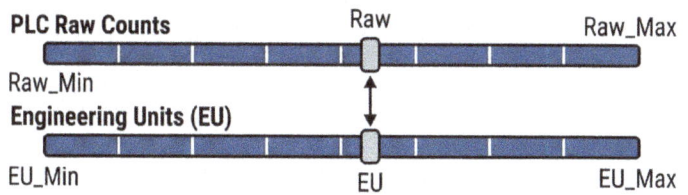

### Example: Rockwell analog input, 0–100 PSI — Worked Example

- **Raw_Min** = 6553 (4 mA)
- **Raw_Max** = 32767 (20 mA)
- **EU_Min** = 0 PSI
- **EU_Max** = 100 PSI
- **Raw** = 19660

EU = (19660 − 6553) ÷ (32767 − 6553) × (100 − 0) + 0
EU = 13107 ÷ 26214 × 100
**EU ≈ 50 PSI**

## Common Raw Count Ranges (4–20 mA)

| Platform | 4 mA | 20 mA |
| --- | --- | --- |
| Rockwell / Allen-Bradley → | 6,553 | 32,767 |
| Siemens S7 → | 5,530 | 27,648 |
| Schneider / Modicon → | 6,400 | 32,000 |

Always verify with specific input card documentation.
Some modules support "raw/proportional" modes and different count ranges.

*APPENDIX A: Pocket Math for the Field*     241

# ELECTRICAL FORMULAS

## Resistors in Series

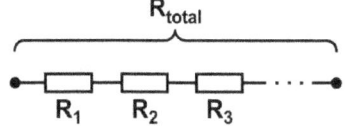

$$R_{total} = R_1 + R_2 + R_3 + \cdots$$

Series resistances add directly

## Resistors in Parallel

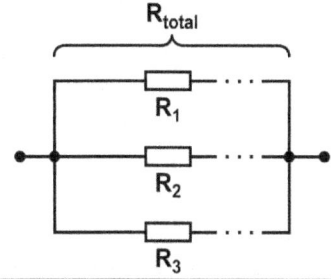

$$\frac{1}{R_{total}} = \frac{1}{R_1} + \frac{1}{R_2} + \frac{1}{R_3} + \cdots$$

Total resistance is always less than the smallest branch resistance

## Two Resistors in Parallel (Shortcut)

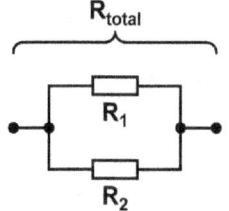

$$R_{total} = (R_1 \times R_2) \div (R_1 + R_2)$$

Valid only for exactly two resistors

## Voltage Divider

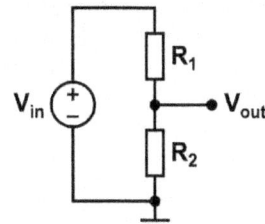

$$V_{out} = V_{in} \times R_2 \div (R_1 + R_2)$$

$R_2$ is the resistor across which $V_{out}$ is measured

Assumes ideal resistors and no load on $V_{out}$

# TIMING

## Frequency to Period

**Period (s) = 1 ÷ Frequency (Hz)**

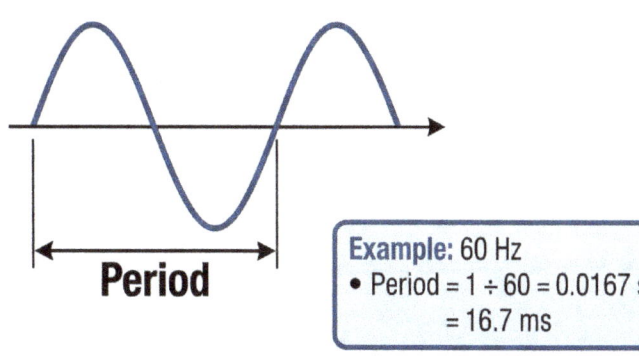

**Example:** 60 Hz
- Period = 1 ÷ 60 = 0.0167 s
  = 16.7 ms

## Pulses to Flow (Pulse Flow Meters)

**Flow = Pulses per unit time ÷ K-factor**

- K-factor = pulses per unit volume (gallon, liter, etc.)

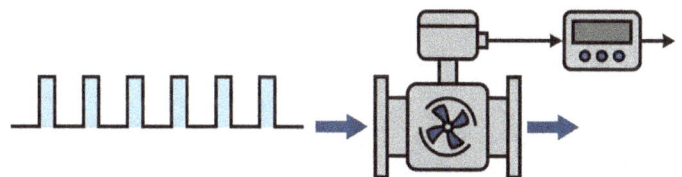

**Example:** 500 pulses in 10 seconds
- Pulses/sec = 500 ÷ 10 = 50 pulses/s
- K-factor = 100 pulses/gallon
- Flow = 50 ÷ 100 = 0.5 gal/s
- Converted result: **30 GPM**

(Vellum-safe) Unit conversions shown elsewhere in Appendix A.

APPENDIX A: *Pocket Math for the Field* 245

## PERCENTAGE ERROR & PERCENT OF SPAN

### 1. Percent Error

$$\% \text{ Error} = \frac{(\text{Measured} - \text{Actual})}{\text{Actual}} \times 100$$

**Example:**

Transmitter reads 52 PSI, test gauge reads 50 PSI

→ (52 − 50) ÷ 50 × 100
= 2 ÷ 50 × 100
= 0.04 × 100
= **4% high**

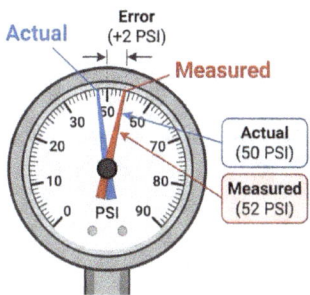

### 2. Percent of Span

$$\% \text{ of Span} = \text{Error} \div \text{Span} \times 100$$

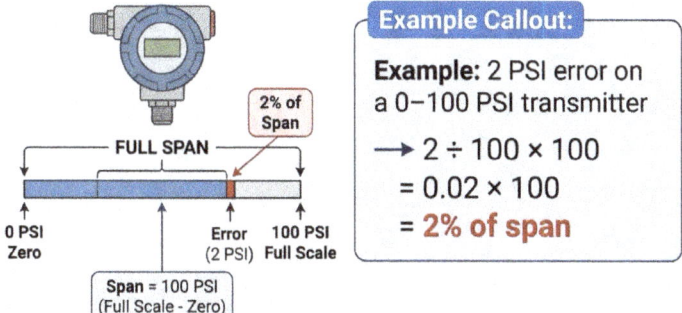

**Example Callout:**

**Example:** 2 PSI error on a 0–100 PSI transmitter

→ 2 ÷ 100 × 100
= 0.02 × 100
= **2% of span**

*"Percent error compares accuracy to the true value. Percent of span compares error to the instrument range."*

*APPENDIX A: Pocket Math for the Field*

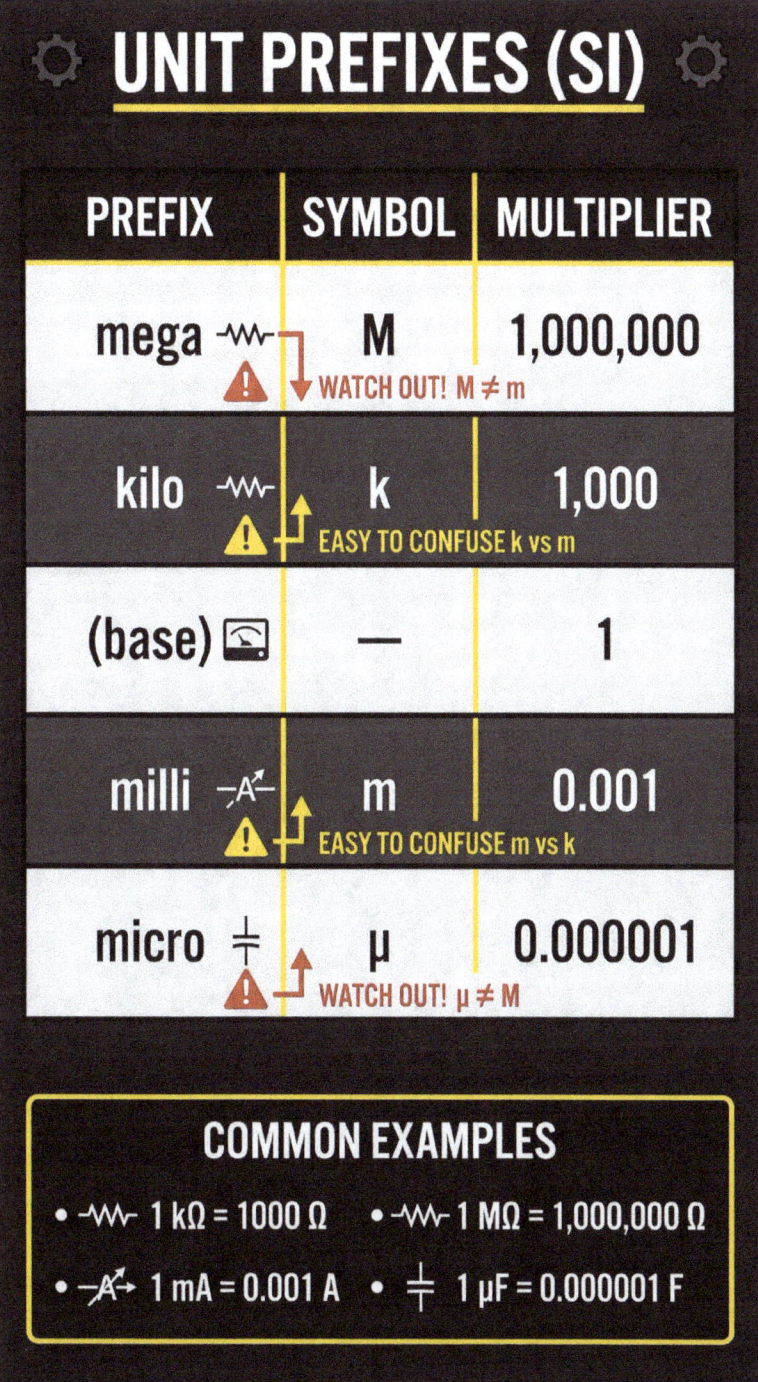

# COMMON FIELD CONVERSIONS

## SECTION 1: LENGTH / DISTANCE

| From | To | Multiply By |
|---|---|---|
| inches → | mm | 25.4 |
| inches → | cm | 2.54 |
| feet → | meters | 0.305 |
| mm → | inches | 0.0394 |
| meters → | feet | 3.281 |

## SECTION 2: VOLUME

| From | To | Multiply By |
|---|---|---|
| gallons (US) → | liters | 3.785 |
| liters → | gallons (US) | 0.264 |
| $ft^3$ → | gallons | 7.48 |
| gallons → | $ft^3$ | 0.134 |

## SECTION 3: WEIGHT / MASS

| From | To | Multiply By |
|---|---|---|
| pounds → | kg | 0.454 |
| kg → | pounds | 2.205 |
| ounces → | grams | 28.35 |
| grams → | ounces | 0.035 |

**Water Weight Reference:**
1 gallon of water ≈ 8.34 lb
≈ 3.78 kg

APPENDIX A: *Pocket Math for the Field* 251

# QUICK MENTAL MATH & FIELD CHECKS

## PERCENTAGE SHORTCUTS

- 10% of X = X ÷ 10
- 1% of X = X ÷ 100
- 25% of X = X ÷ 4
- 50% of X = X ÷ 2

## 4–20 mA QUICK CHECK

- 4 mA = 0%
- 12 mA = 50%
- 20 mA = 100%
- Each 1.6 mA = 10%

## TEMPERATURE QUICK CHECK

- 0°C = 32°F (freezing)
- 100°C = 212°F (boiling)
- 20°C ≈ 68°F (room temperature)

## PRESSURE QUICK CHECK

- 1 PSI ≈ 2.3 ft of water
- 1 ft of water ≈ 0.43 PSI
- 1 bar ≈ 14.5 PSI

**Sanity-check your numbers before trusting the result.**

# APPENDIX B: DISCRETE TROUBLESHOOTING

*These pages are provided to supplement Chapter 11: Discrete Inputs and Chapter 12: Discrete Outputs*

*You are free to photocopy and laminate them for field use.*

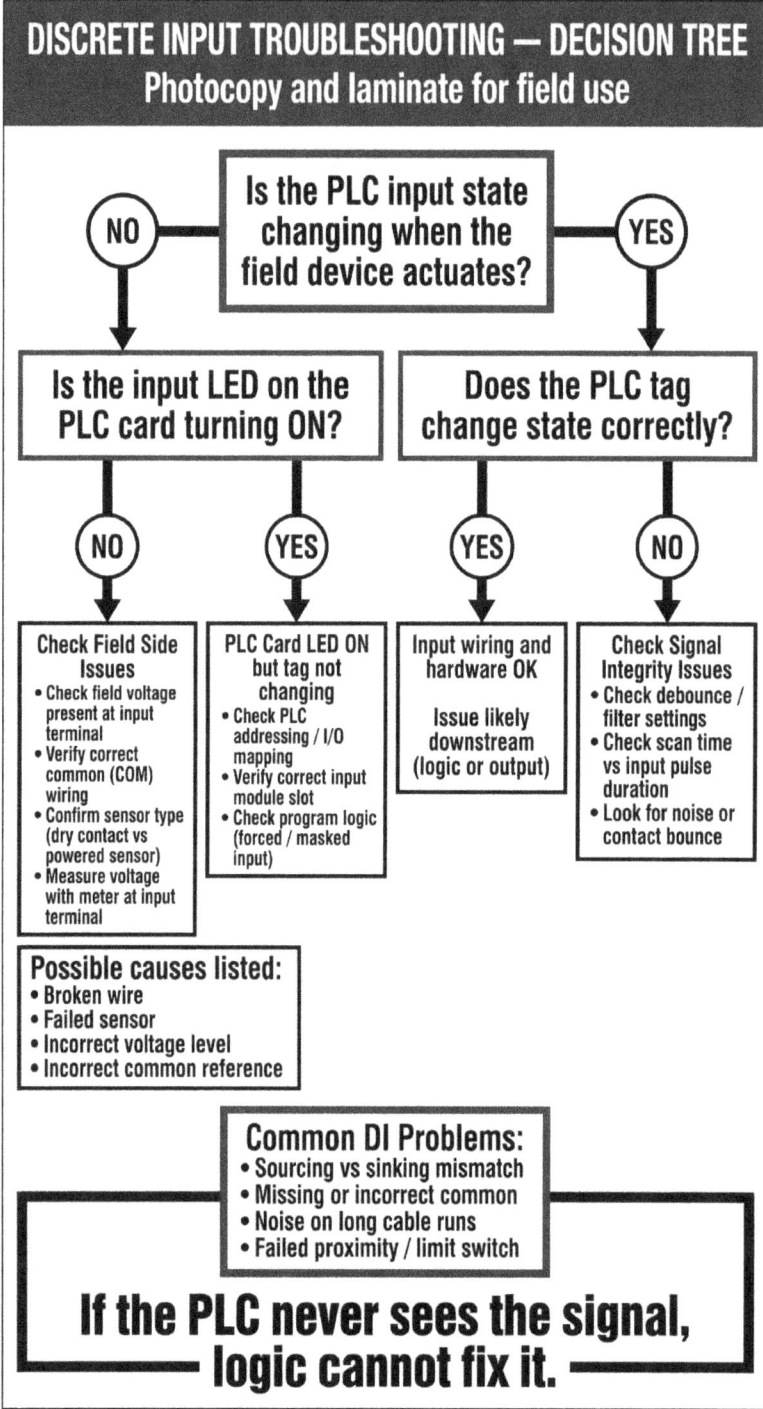

APPENDIX B: Discrete Troubleshooting    255

## PLC DIAGNOSTIC LOOP

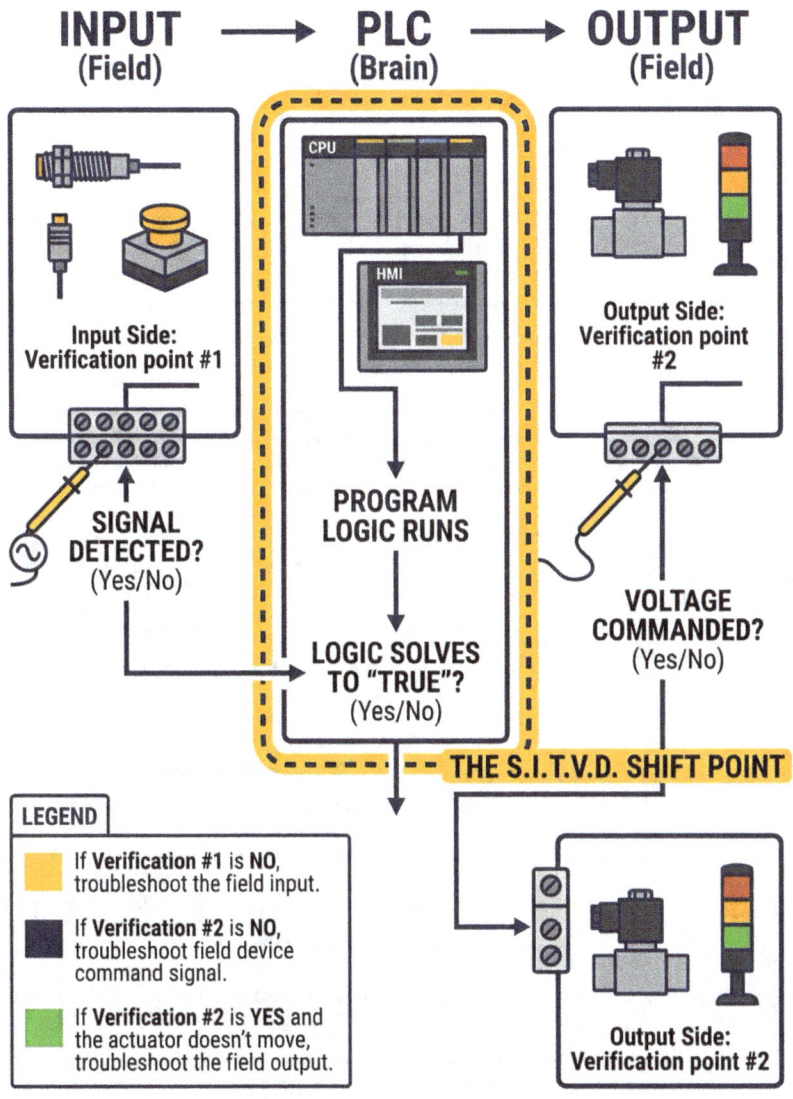

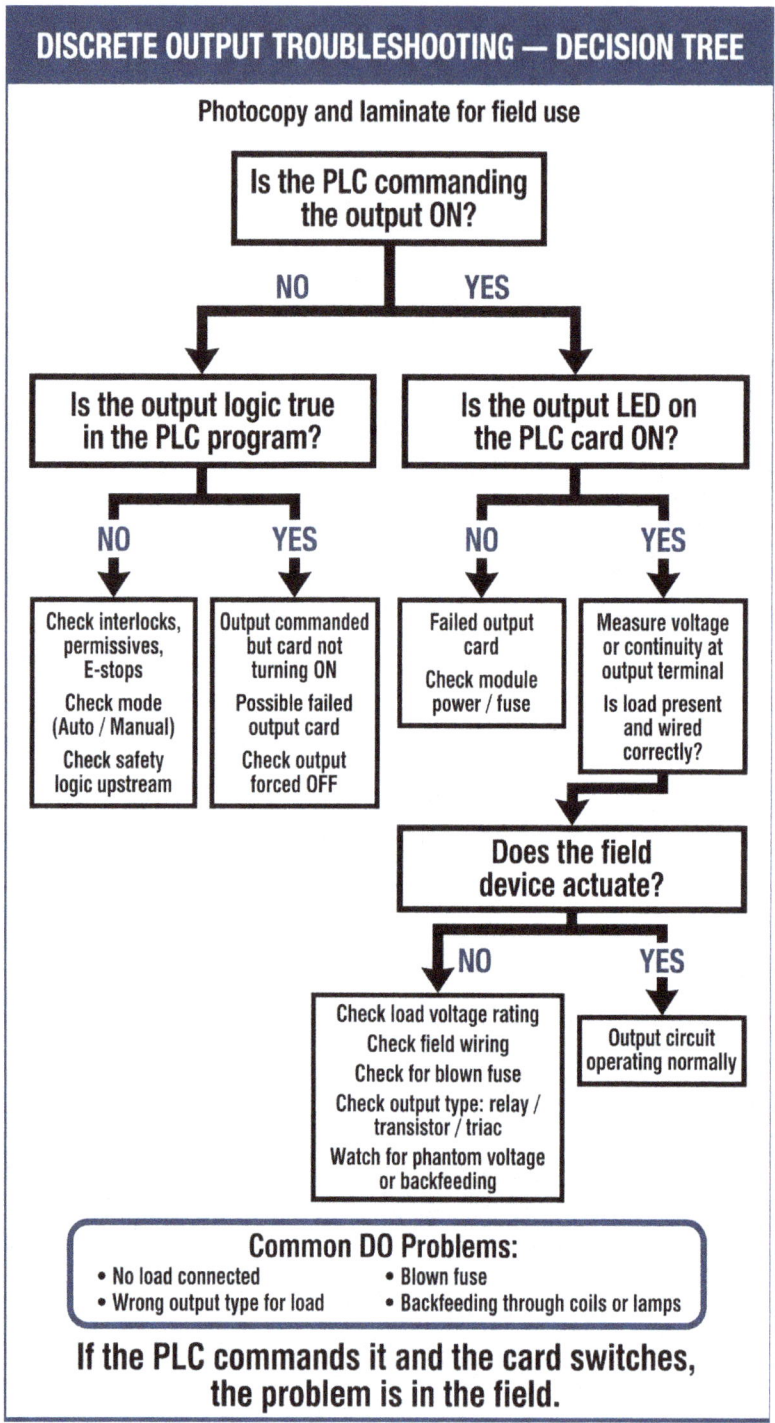

# APPENDIX C: ANALOG LOOP (4-20 MA) TROUBLESHOOTING

*Photocopy the following page and laminate it for field use.*

How to use this flowchart: Start at the top with your symptom—the reading is wrong, missing, or erratic. Follow the decision path: first verify loop power at the card terminals, then measure current in series at the transmitter. If current matches the process, the transmitter is good and the problem is downstream (card, scaling, or display). If current doesn't match, work backward: is the transmitter powered? Is the wiring intact? Is the transmitter configured correctly? Each branch isolates one link in the chain until you find the break.

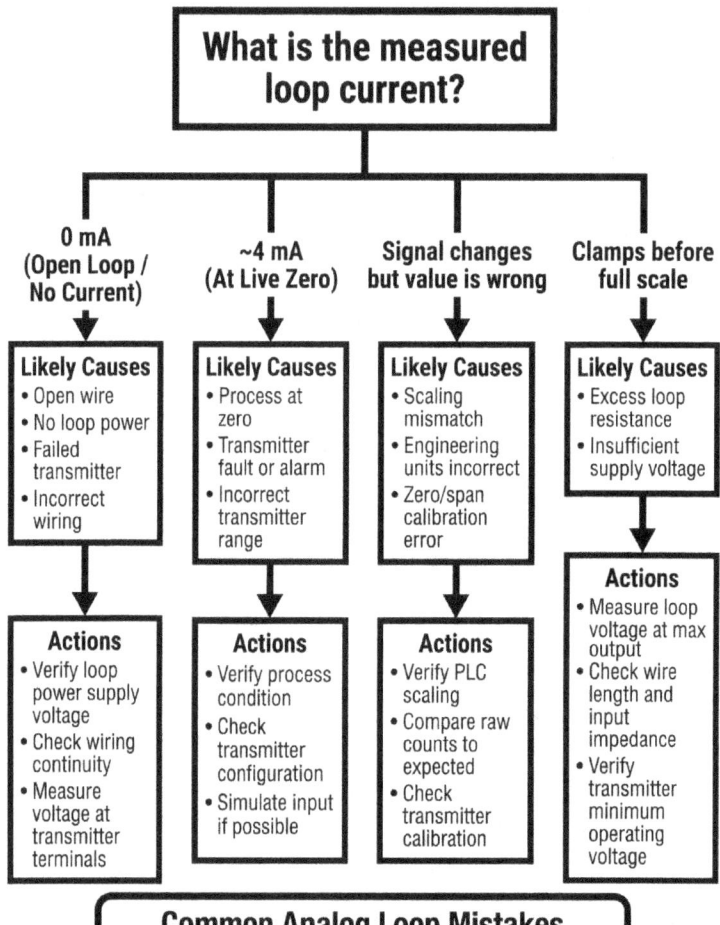

# APPENDIX D: PROXIMITY SENSOR TROUBLESHOOTING

*Photocopy the following page and laminate it for field use.*

How to use this flowchart: Start with the symptom—the sensor isn't detecting, is stuck ON, or is intermittent. First check: is the sensor LED tracking the target? If yes, the sensor is working and the problem is between the sensor output and the PLC (wiring, card, or configuration). If the LED isn't responding, check power, then sensing distance, then target material compatibility. For intermittent faults, focus on mounting rigidity, target consistency, and electrical noise. The chart gets you to the failed component in under five minutes.

260  INSTRUMENTATION & SIGNALS

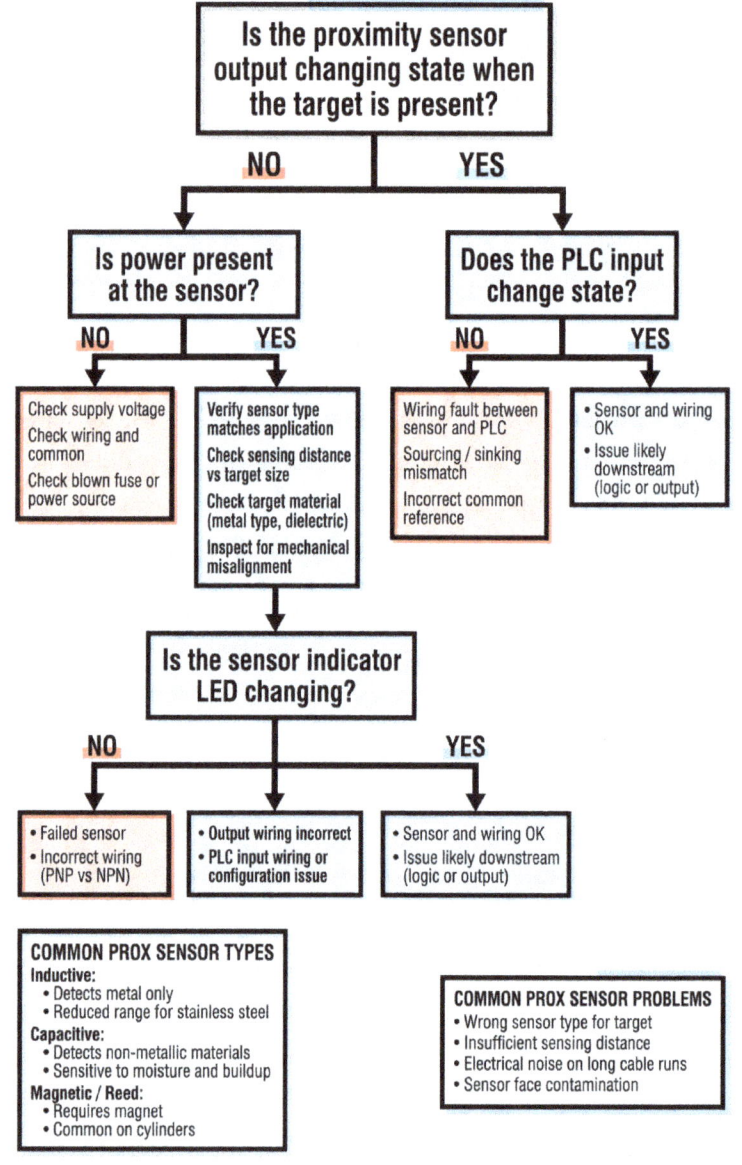

If the sensor never switches locally, the PLC will never see it.

# APPENDIX E: PHOTOEYE TROUBLESHOOTING

*Photocopy the following page and laminate it for field use.*

How to use this flowchart: Start with what you observe—no detect, false detect, or intermittent operation. First verify the indicator LED on the sensor itself. A through-beam sensor with a dead emitter looks identical to a blocked beam—check both ends. For retroreflective sensors, a dirty or missing reflector is the most common failure. For diffuse sensors, background objects and surface reflectivity changes cause most false triggers. The chart walks you through power, alignment, sensitivity adjustment, and output verification in the right order.

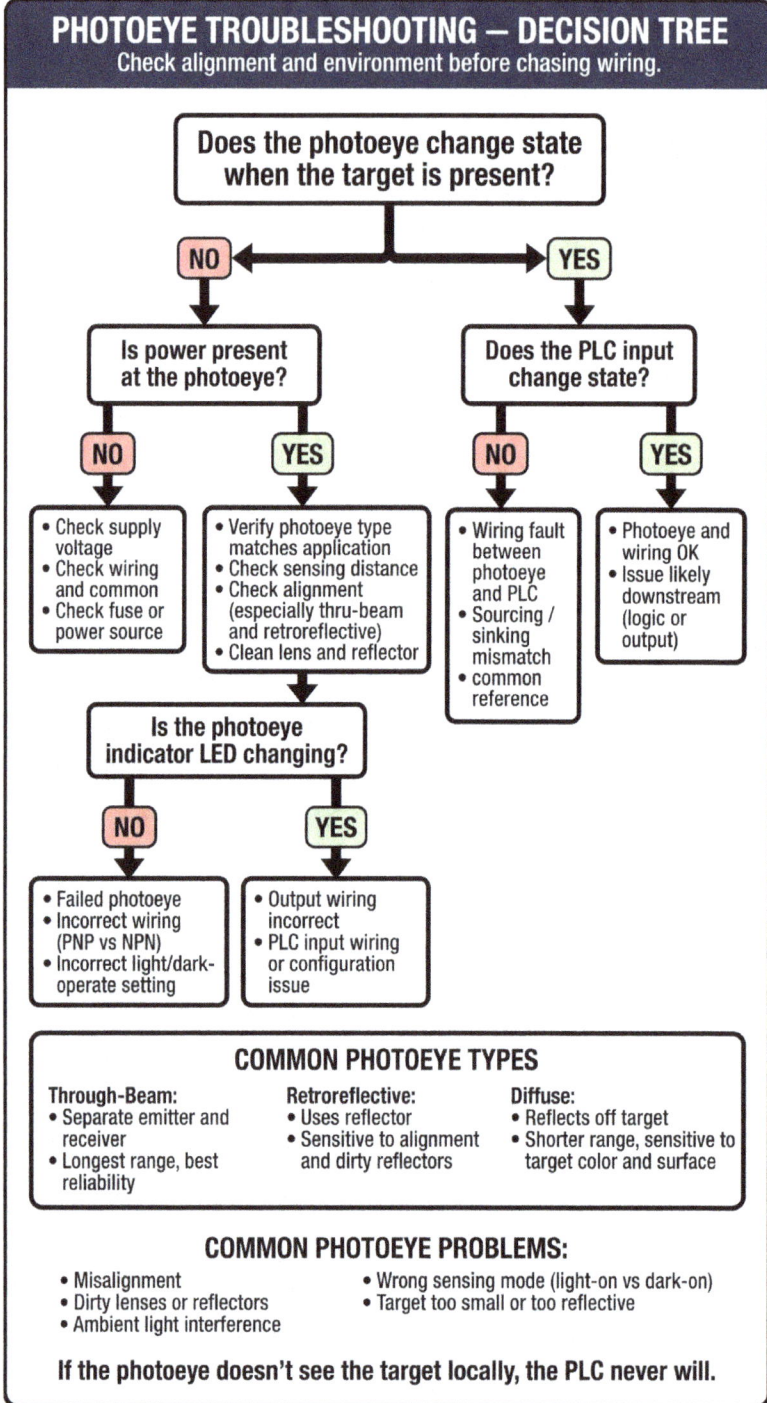

# APPENDIX F: CABLE AND WIRING

*Photocopy these pages and laminate it for field use.*

Cable Type Selection Chart: Page 264

Wire Selection Guide: Page 265

Common Connector Pinouts: Page 266

Cable Trouble Shooting Guide: Page 267

Cable Jacket Decoder: Page 268

Cable Testing Quick Reference: Page 269

Cable and Wiring Troubleshooting Decision: Page 270

# CABLE TYPE SELECTION CHART

| Cable Type | Best Use | Key Features | When NOT to Use |
|---|---|---|---|
| Shielded Twisted Pair (STP) | PLC communications, analog signals, sensor signals in noisy environments | Effective EMI/RFI protection, balanced signal pairs, requires proper grounding | Standard digital inputs/outputs (I/O), low-noise areas (unnecessary cost). |
| Unshielded Twisted Pair (UTP) | Basic digital logic (limit switches, pushbuttons), clean Ethernet connections | Low cost, easier termination, more flexible than shielded | VFD and motor circuits, long runs near high voltage, sensitive analog. |
| Flex-rated Cable | Occasional movement: machine doors, simple articulating arms, pendants | Multi-cycle durability, fine-stranded conductors, durable jacket | Continuous-flex applications (millions of cycles), drag chains, static installations. |
| Fixed Installation Cable (e.g., Tray Cable/TC) | Long-run power, control, signals in cable trays or conduit (static machines) | Robust jacket (often oil/sun resistant), tray rated (TC-ER), simplifies installation | Any repeated flexing or continuous movement, robotics, cable carriers. |
| Armored Cable (MC/AC) | Exposed wiring, physical protection without conduit, power drops to machinery | Metallic armor (steel or aluminum) for crush resistance, robust construction | Tight bend radii, high flexing, highly corrosive environments (check ratings). |
| Conduit-rated Wire (THHN/MTW) | Building discrete circuits inside a conduit system for power or control | Single conductors, moisture & heat resistant, MTW has higher stranding | Outside of conduit (exposed), as a flexible cord, continuous movement. |
| Outdoor/Direct Burial | External sensors, CCTV, remote panels exposed to weather | UV resistant, moisture resistant (e.g., PE or special PVC jacket). | Interior applications (unnecessary), continuous flexing. |
| Oil-Resistant | Machining centers, areas with cutting fluids, hydraulic oils, and lubricants | Jacket compounds designed not to degrade, swell, or crack with common industrial oils | Dry environments (unnecessary cost), where incompatible chemicals are present. |
| High-Flex (Continuous Flexing) | Cable carriers, drag chains, multi-axis robots, moving machine parts | Extremely fine stranding, optimized construction, high wear-resistance, tested for millions of cycles | Static installations, applications with only occasional flexing (excessive cost). |

Refer to local codes and manufacturer specifications for all installations.

# APPENDIX F: Cable and Wiring

## INDUSTRIAL WIRE GAUGE SELECTION GUIDE: DISCRETE MANUFACTURING

| CURRENT CAPACITY: GENERAL GUIDE (18-10 AWG) | | |
|---|---|---|
| AWG | Approx. Max Current (Amps) | Typical Applications (Discrete Mfg) |
| 18 AWG | 7A | Sensor cables, small DC solenoids, pilot lights, standard I/O in packaging equipment. |
| 16 AWG | 10A | Control panel primary wiring, generic 24VDC distribution, medium relays. |
| 14 AWG | 15A | Branch circuits (small motors), heaters, AC lighting, conveyor control power. |
| 12 AWG | 20A | Motor circuits (e.g., small conveyor motors), power feeders, substantial branch circuits for presses. |
| 10 AWG | 30A | Heavy motor branch circuits (e.g., larger pumps, heavy heaters), power distribution for machinery. |

## QUICK VOLTAGE DROP REFERENCE

Keep total voltage drop < 5% from the source for optimal performance.
(e.g., 24VDC - max 1.2V drop, 120VAC - max 6V drop)
**Control/Signal (e.g., 24VDC PLC I/O, sensors):** Aim for < 3-5% drop (e.g., < 1V). Highly sensitive analog signals may need < 2% drop to avoid erratic readings.
**Power Branch Circuits (e.g., 120VAC motors, heaters):** Target < 3% drop within the branch (e.g., max 3.6V for 120VAC) to maintain device efficiency and performance.

## DISTANCE RULES OF THUMB (DISCRETE MFG)

**General Upsizing:** Consider upsizing for runs over 75 feet to mitigate voltage drop, especially with heavier loads.
**24VDC I/O (Sensors, Solenoids):** For common digital I/O like proximity or photoelectric sensors on conveyors or large machines, 18 AWG is typical. If running multi-conductor cables for large packaging machine layouts over 75-100 feet, especially for multiple solenoids (higher current loads), upsizing to 16 AWG helps maintain stable operation.
**120/240/480VAC Power Branches:** Avoid running 14 AWG for motor runs that are long (> 100 ft) and close to full load. Switch to 12 AWG or 10 AWG to minimize voltage drop and prevent reduced motor torque and overheating (e.g., for conveyor drives, press hydraulic pumps).

## CONTROL WIRE VS POWER WIRE SIZING DIFFERENCES

**Control Wire (e.g., PLC signals, sensors, communication):** Primary focus is on signal integrity, managing voltage drop over longer small-current runs, noise rejection (shielding, twisted pairs), and mechanical handling/flexibility (finer strands). Size is often determined by ruggedness and minimizing voltage drop over small current levels for distant devices.
**Power Wire (e.g., motor branches, heaters):** Main concern is current-carrying capacity (ampacity), safety, and heat dissipation. Sizing is primarily based on full-load current calculations, code compliance (e.g., NEC), and heat management for continuous high loads.

## WHY SENSOR CABLES ARE TYPICALLY 18 AWG (Even at Low Current)

**Mechanical Durability:** Sensor wiring in manufacturing environments faces significant vibration, tension during installation, and rough handling around conveyors and presses. 18 AWG provides necessary ruggedness to prevent breaking, even if 24 AWG could electrically carry the low current.
**Reliable Termination:** Larger wire is essential for reliable contact with robust industrial connectors (e.g., M12, M8) and termination methods like crimp terminals or screw clamps on terminal blocks. Fine strands are too fragile and difficult to terminate consistently.
**Signal Stability:** A larger copper cross-section reduces resistance, minimizing voltage drop and ensuring stable, clean 24VDC signal levels, particularly when dealing with small signal variations or multiple sensor signals combining over longer runs.

## M12 A-CODE CONNECTORS (3, 4, 5-PIN)

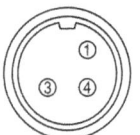  

Key position        Key position        Key position

Pin 1: Brown (+24V)    Pin 1: Brown (+24V)    Pin 1: Brown (+24V)
Pin 3: Blue (0V)       Pin 2: White (Signal 2) Pin 2: White (Signal 2)
Pin 4: Black (Signal)  Pin 3: Blue (0V)       Pin 3: Blue (0V)
                       Pin 4: Black (Signal 1) Pin 4: Black (Signal 1)
                                              Pin 5: Gray (Functional
                                              Earth/Additional Signal)

Key: Single A-coding key slot
Cross-reference to IEC 61076-2-101

## M12 D-CODE 4-PIN (ETHERNET)

D-coding keyway

Pin 1: White/Orange (TX+)
Pin 2: White/Green (RX+)
Pin 3: Orange (TX-)
Pin 4: Green (RX-)

Key: Two D-coding key slots
Cross-reference to IEC 61076-2-101

## M8 CONNECTORS (3, 4-PIN)

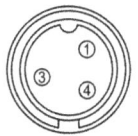 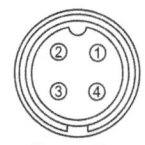

Key position        Key position
**3-pin**           **4-pin**
Pin 1: Brown (+24V)    Pin 1: Brown (+24V)
Pin 3: Blue (0V)       Pin 2: White (Signal 2)
Pin 4: Black (Signal)  Pin 3: Blue (0V)
                       Pin 4: Black (Signal 1)

Key: Single M8 keying slot
Cross-reference to IEC 61076-2-104

## QUICK DISCONNECT (QD) (3, 4-WIRE NEMA)

**3-wire DC style**        **4-wire DC style**
Pin 1: Brown (+24V)        Pin 1: Brown (+24V)
Pin 2: Blue (0V)           Pin 2: White (Signal 2)
Pin 3: Black (Signal)      Pin 3: Blue (0V)
                           Pin 4: Black (Signal 1)

Key: Single key slot
Cross-reference to NEMA standard

## DEVICENET (5-PIN M12/MINI)

Pin 1: Drain (Shield)
Pin 2: V+ (+24V) [Red]
Pin 3: V- (0V) [Black]
Pin 4: CAN H (Signal H) [White]
Pin 5: CAN L (Signal L) [Blue]

Mating face

Key: A-coding key slot (if M12) or equivalent for Mini
Cross-refe. to ODVA Standard

## ETHERNET/IP via M12 D-CODE

Refer to M12 D-code 4-Pin for Pinout & Diagram.

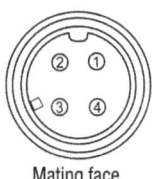

**INDUSTRIAL ETHERNET / ETHERNET/IP**

Pin 1: White/Orange (TX+)
Pin 2: White/Green (RX+)
Pin 3: Orange (TX-)
Pin 4: Green (RX-)

Mating face

Key: D-coding key slots.

APPENDIX F: Cable and Wiring

## INDUSTRIAL CABLE TROUBLESHOOTING GUIDE

### 1. CORE DIAGNOSTIC PROCEDURES

| Test Type | Procedure / Tool | Acceptance Criteria |
|---|---|---|
| Continuity | DMM (Ohms). Power OFF. Isolate both ends. Measure end-to-end on same conductor. | **PASS:** < 2Ω (Length dependent) <br> **FAIL:** > 10Ω or Open Loop (OL) |
| Insulation (IR) | Megohmmeter. 500VDC for 300/600V cable. Test conductor-to-conductor and conductor-to-ground. | **PASS:** > 100 MΩ <br> **MARGINAL:** 2 - 100 MΩ <br> **FAIL:** < 2 MΩ |
| Shield Continuity | DMM (Ohms). Verify shield drain wire is landed at ONE end. Test shield-to-ground at floating end. | **GROUNDED END:** < 1Ω to PE <br> **FLOATING END:** OL to PE |

### 2. TECHNICAL VALIDATION: SCOPE OF TESTS

| Test Type | What it Proves | What it MISSES |
|---|---|---|
| Continuity | Copper is physically intact. | Insulation health or "shorting" risk. |
| Insulation (IR) | Jacket/Dielectric integrity. | Loose terminations (high resistance). |
| Shielding | EMI protection; single-point ground. | Signal quality/crosstalk issues. |

### 3. TROUBLESHOOTING DECISION MATRIX

**Symptom:** Continuity PASS | Insulation FAIL (Low MΩ)
**Probable Cause:** Mechanical damage to jacket, moisture ingress, or crushed insulation.

**Symptom:** Continuity FAIL (High Ω) | Insulation PASS
**Probable Cause:** Fatigue/broken strands at flex point, loose terminal, or poor crimp.

**Symptom:** IR Fails only between Conductors (not Ground)
**Probable Cause:** Internal heat damage, chemical degradation, or conductor-to-conductor arc.

**Symptom:** Shield shows Continuity to Ground at BOTH ends
**Probable Cause:** Ground loop created. Likely to cause signal noise/intermittent errors.

*WARNING: Always discharge cables after Megger testing. Confirm 0V presence before handling. Use Category III/IV rated leads only.*

# INDUSTRIAL CABLE MARKING DECODER

**Practical Example:** `18 AWG 4C STOOW 600V 105°C`
*Decoded:* 18 Gauge, 4 Conductors, Extra Hard Service Portable Cord, Oil Resistant (Jacket & Insulation), Weather/Water Resistant, 600 Volts max, 105°C rated.

## 1. PORTABLE CORD LETTER CODES (SERVICE CORDS)

| | |
|---|---|
| S | Service Grade (600V) |
| J | Junior Service (300V) - Thinner jacket |
| T | Thermoplastic (PVC) |
| E | Elastomer (Rubber-like plastic) |
| O | Oil-Resistant Jacket |
| OO | Oil-Resistant Jacket AND Oil-Resistant Insulation |
| W | Weather/Water Resistant (Outdoor rated) |

## 2. BUILDING & MACHINE TOOL WIRE (FIXED)

| | |
|---|---|
| MTW | Machine Tool Wire (Stranded, flexible for panels, moisture resistant) |
| THHN | Thermoplastic High Heat-resistant Nylon coated (Dry locations) |
| THWN | Thermoplastic High Water-resistant Nylon coated (Wet locations) |
| TC / PLTC | Tray Cable / Power Limited Tray Cable (For cable trays/raceways) |

## 3. PHYSICAL & ENVIRONMENTAL RATINGS

| | |
|---|---|
| Temp Range | Standard is 60°C to 105°C. Lower limits (e.g., -40°C) indicate cold-bend/impact suitability. |
| Voltage | 600V (Standard Industrial); 300V (Junior/Control/Signal). |
| Flex Rating | **Continuous Flex:** Rated for millions of cycles in C-tracks/motion. **Flexible/Limited Flex:** Designed for easy routing/occasional movement only. |
| Compliance | **UL:** Underwriters Laboratories (US). **CSA:** Canadian Standards Assoc. |

## 4. CONDUCTOR FORMAT

[Gauge] AWG [Count]C: e.g., `14 AWG 3C` = 14 gauge wire with 3 conductors.
**G / W-G:** Indicates a dedicated Ground wire is included (e.g., 12/3 G).

---

Reference for Industrial Maintenance & Field Engineering. Adhere to NEC/NFPA 70 guidelines for installation.

*APPENDIX F: Cable and Wiring* 269

## INDUSTRIAL CABLE TROUBLESHOOTING GUIDE

### 1. TESTING PROCEDURES & CRITERIA

| Test Type | Procedure / Settings | Acceptance Criteria |
|---|---|---|
| **Continuity** (Loop Test) | DMM (Ohms). Isolate both ends. Short conductors at far end; measure at near end. | **GOOD:** < 2.0 Ω <br> **SUSPECT:** 2.0 Ω - 10 Ω <br> **FAIL:** > 10 Ω or OL |
| **Insulation** (IR/Megger) | Megohmmeter. Set to 500VDC (600V cable). Test conductor-to-conductor and cond-to-ground. | **GOOD:** > 100 MΩ <br> **MARGINAL:** 2 - 100 MΩ <br> **FAIL:** < 2 MΩ |
| **Shield Integrity** | DMM (Ohms). Verify drain wire landed at source end. Measure shield-to-ground at load end. | **SOURCE:** < 1.0 Ω to PE <br> **LOAD (Floating):** OL to PE |

### 2. TECHNICAL VALIDATION MATRIX

| Test Type | Proves (Validation) | Does NOT Prove |
|---|---|---|
| Continuity | Path is physically closed. | Insulation health; load capacity. |
| Insulation | Dielectric barrier is intact. | Good terminations/crimps. |
| Shielding | Noise rejection; no ground loop. | Signal frequency compliance. |

### 3. TROUBLESHOOTING DECISION MATRIX

**IF** Continuity PASS | Insulation FAIL:
**THEN** Inspect for jacket breach, moisture ingress, or internal carbon tracking.

**IF** Continuity FAIL | Insulation PASS:
**THEN** Inspect for fatigue at flex points, loose terminal screws, or oxidized crimps.

**IF** Shield to Ground < 100 Ω at Floating End:
**THEN** Ground loop exists. Locate contact point where shield touches metal/conduit.

**IF** Insulation Fails Cond-to-Cond Only:
**THEN** Cable likely overheated or crushed; jacket may appear normal.

**SAFETY:** Ensure cable is de-energized and locked out. Discharge cable conductors to ground after Megger testing to bleed residual capacitive charge.

270  INSTRUMENTATION & SIGNALS

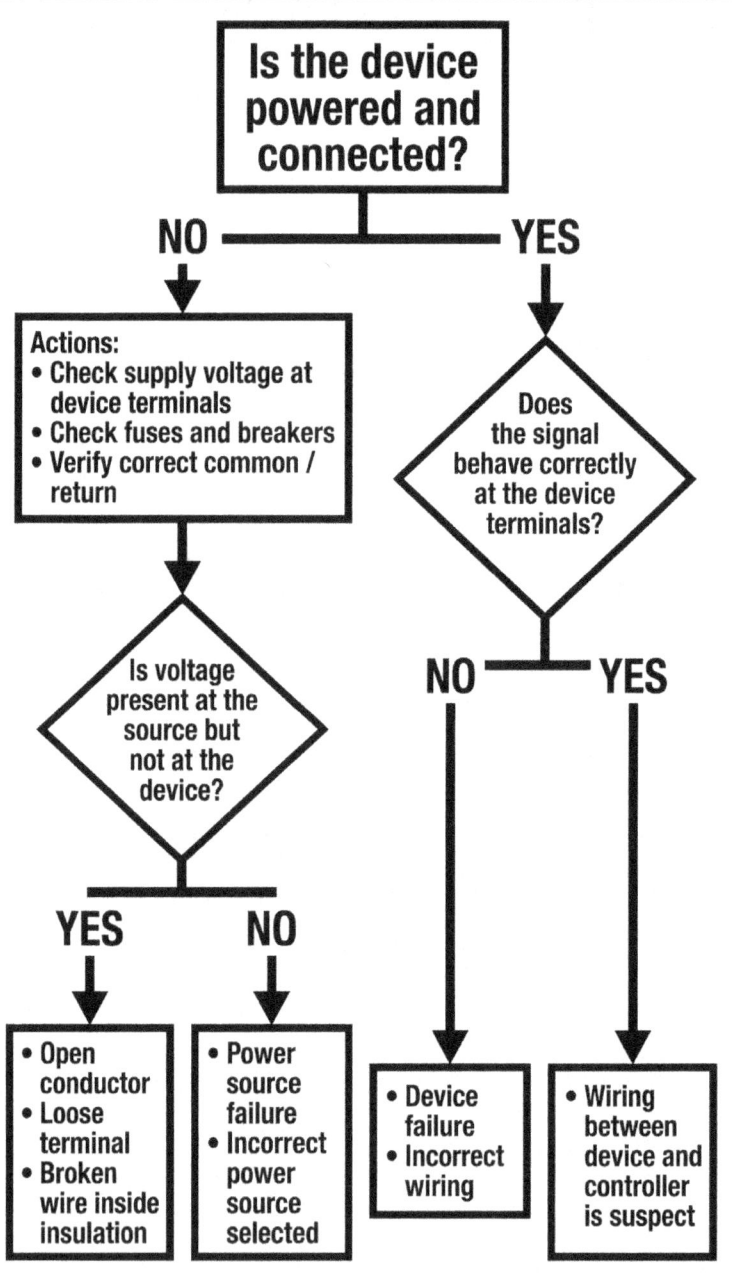

# APPENDIX G: TEMPERATURE SENSOR TROUBLESHOOTING

Temperature sensors fail differently than discrete I/O. An RTD with a broken wire looks like an open circuit. A thermocouple with reversed polarity gives you a believable reading—just the wrong one. An IR sensor with the wrong emissivity setting reads consistently low, and you'll waste hours calibrating a sensor that's working perfectly.

This appendix gives you the reference tables and troubleshooting guides to isolate temperature sensor problems in minutes instead of hours.

**What's included:**

- **RTD Resistance Tables** (pages 273-274): Pt100 and Pt1000 resistance values at common temperatures. Use these to verify the sensor element before you blame the transmitter or card.
- **Thermocouple Voltage Tables** (pages 275-): Type K, J, and T voltage references. Measure millivolts at the junction, compare to the table, and you know instantly whether the thermocouple is good.
- **Sensor Selection Matrix** (page 275-277): Quick reference for choosing the right sensor type for your application—RTD vs. thermocouple vs. IR vs. thermistor.
- **Common Failure Modes** (pages 279-281): Symptom-based troubleshooting tables for RTDs, thermocouples, IR sensors, and thermistors. Most "sensor failures" are actually configuration errors or wiring problems.

- **RTD Wiring Configurations** (pages 282-283): 2-wire vs. 3-wire vs. 4-wire comparison. If you're troubleshooting an RTD and don't understand lead resistance compensation, you'll chase your tail for hours.
- **Temperature Troubleshooting Flowchart** (page 284): Decision tree for systematic troubleshooting. Photocopy this page and laminate it for field use.

**For detailed temperature sensor theory and SITVD troubleshooting workflows, see Chapter 14 (pages 114-121).**

*APPENDIX G: Temperature Sensor Troubleshooting* 273

## RTD RESISTANCE REFERENCE TABLES

These tables show the resistance values for Pt100 and Pt1000 RTDs at common temperatures. Use these for field verification: if you measure resistance at a known temperature and it matches the table, the RTD element is good. If it doesn't match, the RTD is failed or you have lead resistance errors.

### Pt100 RTD (IEC 60751 Standard)

Resistance at 0°C: 100.00 Ω

| Temp (°C) | Temp (°F) | Resistance (Ω) |
|---|---|---|
| -50 | -58 | 80.31 |
| -25 | -13 | 90.19 |
| 0 | 32 | 100.00 |
| 25 | 77 | 109.73 |
| 50 | 122 | 119.40 |
| 75 | 167 | 129.00 |
| 100 | 212 | 138.51 |
| 125 | 257 | 147.95 |
| 150 | 302 | 157.33 |
| 175 | 347 | 166.63 |
| 200 | 392 | 175.86 |
| 225 | 437 | 185.01 |
| 250 | 482 | 194.10 |
| 300 | 572 | 212.05 |
| 350 | 662 | 229.72 |
| 400 | 752 | 247.09 |

**Field Verification Example:** Room temperature is 25°C (77°F). Measure RTD resistance directly at the sensor terminals. You should read approximately 109.7 Ω. If you read 100 Ω, the sensor is cold or failed. If you read 150 Ω, you're measuring lead resistance (3-wire RTD misconfigured as 2-wire) or the sensor is hot.

## Pt1000 RTD (IEC 60751 Standard)

Resistance at 0°C: 1000.0 Ω

| Temp (°C) | Temp (°F) | Resistance (Ω) |
|---|---|---|
| -50 | -58 | 803.1 |
| -25 | -13 | 901.9 |
| 0 | 32 | 1000.0 |
| 25 | 77 | 1097.3 |
| 50 | 122 | 1194.0 |
| 75 | 167 | 1290.0 |
| 100 | 212 | 1385.1 |
| 125 | 257 | 1479.5 |
| 150 | 302 | 1573.3 |
| 175 | 347 | 1666.3 |
| 200 | 392 | 1758.6 |
| 225 | 437 | 1850.1 |
| 250 | 482 | 1941.0 |
| 300 | 572 | 2120.5 |
| 350 | 662 | 2297.2 |
| 400 | 752 | 2470.9 |

**The 10:1 Rule:** Pt1000 resistance is exactly 10 times Pt100 resistance at the same temperature. If you're troubleshooting a Pt1000 and measure 1385 Ω, that's 100°C. The same reading on a Pt100 would be way over-range—if the HMI shows a believable temperature but you measure 1385 Ω on what's supposed to be a Pt100, the input card is configured for the wrong sensor type.

**Critical Note:** These values assume the RTD follows the IEC 60751 standard (Alpha = 0.00385). Most industrial RTDs do. If you encounter an RTD with a different alpha coefficient, consult the manufacturer's data.

APPENDIX G: *Temperature Sensor Troubleshooting* 275

# THERMOCOUPLE VOLTAGE REFERENCE TABLES

Thermocouples generate millivolts (mV) proportional to temperature. These tables show the voltage you should measure at the thermocouple junction for common types. All values assume a 0°C (32°F) reference junction (cold junction compensation).

**Critical:** Thermocouple voltages are small—measured in millivolts. If you measure volts instead of millivolts, you're not measuring the thermocouple signal.

## Type K Thermocouple (Chromel-Alumel)

Most common general-purpose thermocouple. Temperature range: -200°C to +1372°C.

| Temp (°C) | Temp (°F) | Voltage (mV) |
|---|---|---|
| -200 | -328 | -5.891 |
| -150 | -238 | -4.913 |
| -100 | -148 | -3.554 |
| -50 | -58 | -1.889 |
| 0 | 32 | 0.000 |
| 50 | 122 | 2.023 |
| 100 | 212 | 4.096 |
| 150 | 302 | 6.138 |
| 200 | 392 | 8.138 |
| 250 | 482 | 10.153 |
| 300 | 572 | 12.209 |
| 400 | 752 | 16.397 |
| 500 | 932 | 20.644 |
| 600 | 1112 | 24.905 |
| 800 | 1472 | 33.275 |
| 1000 | 1832 | 41.276 |
| 1200 | 2192 | 48.838 |

**Typical Use:** General purpose, industrial furnaces, kilns, engine exhaust

## Type J Thermocouple (Iron-Constantan)

Lower temperature range. Common in older equipment. Temperature range: -210°C to +1200°C.

| Temp (°C) | Temp (°F) | Voltage (mV) |
|---|---|---|
| -200 | -328 | -7.890 |
| -150 | -238 | -6.500 |
| -100 | -148 | -4.633 |
| -50 | -58 | -2.431 |
| 0 | 32 | 0.000 |
| 50 | 122 | 2.585 |
| 100 | 212 | 5.269 |
| 150 | 302 | 8.010 |
| 200 | 392 | 10.779 |
| 250 | 482 | 13.555 |
| 300 | 572 | 16.327 |
| 400 | 752 | 21.846 |
| 500 | 932 | 27.393 |
| 600 | 1112 | 33.102 |
| 700 | 1292 | 39.130 |
| 800 | 1472 | 45.494 |

**Typical Use:** Plastics extrusion, lower-temperature industrial processes

APPENDIX G: *Temperature Sensor Troubleshooting* 277

## Type T Thermocouple (Copper-Constantan)

Best for low and cryogenic temperatures. Temperature range: -270°C to +400°C.

| Temp (°C) | Temp (°F) | Voltage (mV) |
|---|---|---|
| -200 | -328 | -5.603 |
| -150 | -238 | -4.648 |
| -100 | -148 | -3.379 |
| -50 | -58 | -1.819 |
| 0 | 32 | 0.000 |
| 50 | 122 | 2.036 |
| 100 | 212 | 4.279 |
| 150 | 302 | 6.704 |
| 200 | 392 | 9.288 |
| 250 | 482 | 12.013 |
| 300 | 572 | 14.862 |
| 350 | 662 | 17.819 |
| 400 | 752 | 20.872 |

**Typical Use:** Cryogenic applications, refrigeration, food processing

**Field Verification:**

1. Disconnect the thermocouple from the input card
2. Set your meter to mV DC
3. Measure voltage directly at the thermocouple junction
4. Compare to the table at the known process temperature
5. If voltage matches, the thermocouple is good; problem is downstream (wiring, card, or configuration)
6. If voltage doesn't match, the thermocouple is failed or the process temperature is not what you think it is

**Common Mistake:** Measuring with the wrong polarity. Thermocouples have polarity. Reversed leads give you negative readings when you should see positive. If your reading is correct magnitude but wrong sign, swap the leads.

# TEMPERATURE SELECTION MATRIX

## TEMPERATURE SENSOR SELECTION MATRIX

| Selection Criteria | RTD (Pt100) | Thermocouple | Infrared |
|---|---|---|---|
| Temperature Range | −200°C to +850°C (−328°F to +1562°F) | −270°C to +1800°C Type dependent | −50°C to +3000°C Model dependent |
| Accuracy | ±0.1°C (4-wire) ±0.5°C (3-wire) | ±1.0–2.5°C Type dependent | ±1–2% of reading Emissivity dependent |
| Response Time | Slow (seconds) Thermal mass of element | Fast (milliseconds) Small thermal mass | Very fast (ms) No contact required |
| Signal Type | Resistance change (ohms) | Voltage (millivolts) Self-generated | Analog (4–20 mA) or digital output |
| Wiring | 2, 3, or 4-wire 3-wire most common | 2-wire Must match TC type | Power + signal Varies by model |
| Noise Susceptibility | Low (resistance measurement) | High (millivolt signal) | Medium (optical interference) |
| Cost | Medium-High 50–500 | Low 10–100 | High 200–2000+ |
| Best For | Accuracy-critical process control | Wide range, rugged general purpose | Moving targets non-contact required |
| Common Applications | Food processing Pharmaceutical, HVAC | Ovens, furnaces Extruders, heat treat | Conveyors, rotating Hazardous environments |
| Watch Out For | Lead wire resistance (use 3 or 4-wire) | Wrong type/polarity = believable but wrong | Emissivity mismatch Dirty lens |
| Failure Signature | Reads too high (open) or erratic (bad lead) | Reads ambient (open) or offset (wrong type) | Wrong reading (dirty lens or wrong ε) |
| Field Verification | Meter resistance ~110Ω at 25°C (Pt100) | Ice bath = 32°F Boiling water = 212°F | Compare to contact sensor on same surface |

*APPENDIX G: Temperature Sensor Troubleshooting*

# COMMON TEMPERATURE SENSOR FAILURE MODES

## RTD Failures

| Symptom | Probable Cause | Fix |
|---|---|---|
| Reading stuck at min/max | Open circuit (broken element) | Replace RTD |
| Reading consistently high | Lead resistance error | Reconfigure as 3-wire or replace cable |
| Reading drifts over time | Element degradation | Replace RTD; investigate contamination |
| Reading jumps erratically | Intermittent connection | Repair connection or replace cable |
| Reading 10x too high/low | Wrong sensor type (Pt100 vs Pt1000) | Reconfigure card for correct type |
| Stable but offset | Wrong # wires configured | Match card config to wiring |

**RTD Truth:** Most "RTD failures" are wiring problems, not sensor failures. Check wiring configuration and lead resistance before replacing the RTD.

## Thermocouple Failures

| Symptom | Probable Cause | Fix |
|---|---|---|
| Reading at room temp (process hot) | Open junction (broken wire) | Replace thermocouple |
| Reading inverted (negative) | Reversed polarity (leads swapped) | Swap leads |
| Reading offset by fixed amount | Wrong TC type configured | Reconfigure card for correct type |
| Erratic/noisy reading | Ground loop, EMI, or near power | Reroute cable; verify single ground |
| Reading drifts low over months | Junction degradation | Replace thermocouple |
| Reading freezes at one value | Shorted TC (wires touching) | Replace thermocouple |

**Thermocouple Truth:** Thermocouples fail at the junction. If voltage is wrong at the sensing end, the thermocouple is bad. If voltage is correct at the sensor but wrong at the card, the problem is wiring or card configuration.

APPENDIX G: Temperature Sensor Troubleshooting    281

## IR (Infrared) Non-Contact Sensor Failures

| Symptom | Probable Cause | Fix |
|---|---|---|
| Reading 50-100F low on metal | Emissivity setting too high | Set emissivity to match target |
| Reading drifts low over time | Lens contamination (dust, oil) | Clean lens; install air purge |
| No reading or intermittent | Blocked line of sight or no power | Clear obstruction; verify power |
| Reading jumps with background | Spot size larger than target | Move closer or use smaller spot |
| Fluctuates rapidly | Reflections or steam in path | Shield from reflections; clear atmosphere |

**IR Truth:** 90% of IR sensor problems are emissivity mismatches or dirty lenses. The sensor itself rarely fails. Always verify emissivity setting before blaming the sensor.

## NTC Thermistor Failures

| Symptom | Probable Cause | Fix |
|---|---|---|
| Resistance does not change with temp | Failed thermistor element | Replace thermistor |
| Reading way off | Wrong thermistor curve configured | Replace with correct part or reconfigure |
| Erratic reading | Moisture intrusion | Replace thermistor; improve sealing |

**NTC Truth:** Thermistors are used in HVAC and consumer equipment, not heavy industrial. If you find one in a plant, it's probably on an OEM skid. Replace it with the exact OEM part number—thermistors are not interchangeable.

# RTD WIRING CONFIGURATIONS

### 2-Wire vs. 3-Wire vs. 4-Wire

RTDs measure temperature by resistance. The problem: the wires connecting the RTD to the input card also have resistance. Lead resistance adds to the RTD resistance, making the reading appear hotter than reality. Different wiring configurations compensate for this error.

### 2-Wire RTD

**How it works:** The simplest configuration. Two wires connect the RTD to the input card. Current flows out one wire, through the RTD, and back the other wire. The card measures total resistance: RTD + both wire resistances.

**Error:** Lead resistance adds directly to the reading.

- Wire resistance: 1 $\Omega$ per wire = 2 $\Omega$ total
- Pt100 at 25°C = 109.7 $\Omega$
- Measured resistance = 111.7 $\Omega$ → reads as ~30°C instead of 25°C

**When to use:** Very short wire runs (<10 feet), or when accuracy doesn't matter.

**When NOT to use:** Any industrial application with more than 10 feet of wire.

### 3-Wire RTD

**How it works:** The most common industrial configuration. Three wires connect the RTD to the input card. Two wires go to one side of the RTD, one wire goes to the other side. The card measures lead resistance separately and subtracts it mathematically.

**Error:** Lead resistance is mostly compensated. Remaining error is due to slight differences between wire resistances.

- Typical error with 100 feet of cable: ±0.1°C to ±0.3°

**When to use:** Standard for industrial installations. Balances accuracy and cost.

**When NOT to use:** When you need absolute precision (<0.1°C), use 4-wire instead.

# APPENDIX G: Temperature Sensor Troubleshooting

## 4-Wire RTD

**How it works:** The most accurate configuration. Four wires connect the RTD to the input card. Two wires carry excitation current, two separate wires measure voltage. Because the measurement wires carry no current, they have no voltage drop. Lead resistance is completely eliminated.

**Error:** Zero lead resistance error (assuming wires are intact).

• Typical error with 1000 feet of cable: 0.01°C

**When to use:** Precision applications, calibration standards, very long cable runs.

**When NOT to use:** Most industrial applications don't need this level of accuracy and the extra wiring cost isn't justified.

## Comparison Summary

| Config | Wires | Compensation | Error (100 ft) | Cost | Best Use |
|---|---|---|---|---|---|
| 2-Wire | 2 | None | ±2°C to ±5°C | Lowest | Very short runs only |
| 3-Wire | 3 | Good | ±0.1°C to ±0.3°C | Moderate | Standard industrial |
| 4-Wire | 4 | Perfect | <0.01°C | Highest | Precision/ calibration |

**Field Reality:**

Most industrial installations use 3-wire RTDs. If you see a 2-wire RTD on a long run, someone made a mistake—the reading will be wrong and you'll chase your tail trying to calibrate it. If you see a 4-wire RTD in a typical plant application, someone over-specified it.

**Critical Mistake:** Configuring the input card for 3-wire when the RTD is actually wired 2-wire (or vice versa). The card will attempt to compensate for lead resistance that isn't compensated in the wiring, giving you a wrong reading. **Always match the card configuration to the actual wiring.**

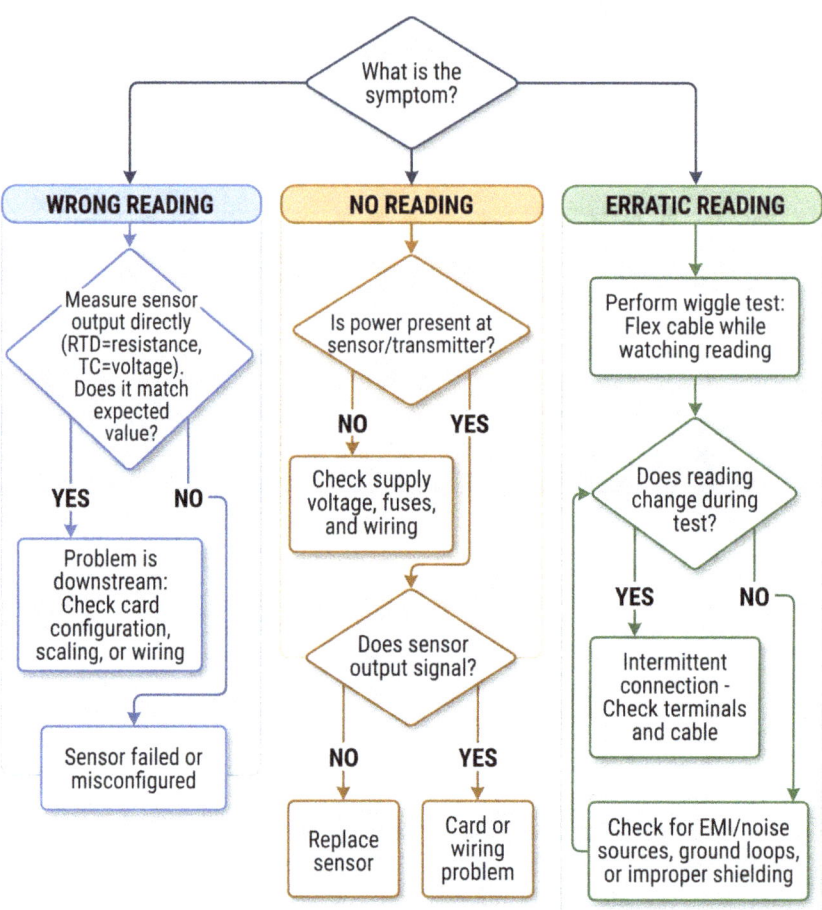

# APPENDIX H: THE TECHNICIAN'S FIELD KIT

Every troubleshooting workflow in this book assumes you have the right tool in your hand. This appendix covers the instruments you need to carry, what each one actually does, and the one mistake most technicians make with each of them.

You don't need every tool on this list to get started. You need the multimeter. Everything else extends your reach.

## THE SIX TOOLS

| Tool | Primary Use | When You Reach For It |
|---|---|---|
| Digital Multimeter (DMM) | Voltage, current, resistance, continuity | Always. |
| Loop Calibrator | Source and simulate 4-20 mA | Any analog loop fault |
| HART Communicator | Talk to smart transmitters | Analog loop configuration, diagnostics |
| Clamp-On mA Meter | Measure loop current non-invasively | When you can't break the loop |
| Insulation Resistance Tester | Megohm resistance — cable and motor health | Moisture damage, insulation breakdown |
| Laptop with PLC Software | Go online, monitor logic, read diagnostics | Any PLC, I/O, or logic fault |

## TOOL 1: DIGITAL MULTIMETER (DMM)

The multimeter is your first tool on every call. Before you touch anything else, the meter tells you whether power is present, whether a signal is there, and whether a circuit is open or closed. A technician without a meter is guessing.

## WHAT TO BUY

Not all meters are equal. Two specifications matter above all others:

### True RMS (Root Mean Square)

Standard average-responding meters measure AC voltage by assuming a perfect sine wave and calculating RMS from the peak. In industrial environments—where VFDs, switching power supplies, and motor loads distort the waveform—average-responding meters provide inaccurate data. True RMS meters measure the actual waveform regardless of shape; for professional industrial troubleshooting, True RMS is not optional.

### CAT (Category) Rating

The CAT rating defines a meter's ability to withstand transient voltage spikes in specific electrical environments. Higher CAT numbers indicate safer operation in higher-energy circuits.

| CAT Rating | Environment | Typical Application |
|---|---|---|
| CAT I | Electronic / signal level | Low-energy circuits only |
| CAT II | Household / receptacle | Single-phase loads at outlets |
| CAT III | Distribution / panel | Fixed equipment, branch circuits, switchgear |
| CAT IV | Utility / service entrance | Overhead lines, utility connections |

## CRITICAL SAFETY NOTE

For general industrial troubleshooting, **CAT III is the minimum** requirement, while **CAT IV is preferred**. If you are working near motor control centers, distribution panels, or anything above **480V**, CAT IV is mandatory. A meter not rated for its environment can fail explosively during a transient event.

## MEASUREMENT MODES AND WHEN TO USE THEM

| Mode & Symbol | Primary Use Cases | Common Mistakes |
|---|---|---|
| DC Voltage (VDC or V⎓) | 24 VDC control power, PLC I/O signals, transmitter terminals. | Measuring AC voltage on DC range; will result in a near-zero reading. |
| AC Voltage (VAC or V~) | 120/240 VAC loads, relay coil power. | Forgetting to switch from DC; will result in a near-zero reading. |
| DC Milliamps (mA⎓) | 4-20 mA loop current (must be measured in series). | Connecting in parallel; blows the fuse and may damage the loop. |
| Resistance (Ω) | Continuity, RTD/coil resistance, and cable integrity. | Measuring on a live circuit; damages the meter and gives wrong readings. |
| Continuity (♪) | Quick wire integrity checks. | Relying on it for quality checks; it doesn't confirm proper termination. |
| Diode Test | Checking suppression diodes and transistor output verification. | None specified. |

⚠ CRITICAL: Never connect your meter in milliamp (mA) mode across a voltage source. The milliamp input is protected by a fuse—but a high-energy source can blow the fuse before the protection acts, damage the meter, or create a dangerous arc.

## MULTIMETER LEAD PLACEMENT

- **Voltage measurements:** leads connect in parallel with the circuit. Current never flows through your meter on a voltage measurement.
- **Current measurements (mA mode):** leads connect in series— **break** the circuit and insert the meter into the current path. The loop current flows through your meter.
- **Resistance measurements:** circuit must be de-energized. Measuring resistance on a live circuit gives wrong readings and damages the meter.

## THE GHOST VOLTAGE PROBLEM

High-impedance meters (most modern DMMs have 10 MΩ input impedance) pick up induced voltages on unloaded conductors. You can read 15 VDC on a wire that is not connected to anything—because the wire is running alongside a 24 VDC cable and capacitively coupling the voltage. This is ghost voltage. It looks real. It will mislead you.

**The Verification Test**

To confirm if a reading is ghost voltage, connect a **known load** (such as a 10 kΩ resistor or an indicator lamp) across the measurement points. Ghost voltage will collapse immediately under the load, whereas real voltage will hold.

## TOOL 2: LOOP CALIBRATOR

The loop calibrator is the multimeter's partner for analog instrumentation work. Where the multimeter reads, the calibrator sources and simulates. Together, these two tools can isolate any 4-20 mA loop fault to a specific component in the chain.

### WHAT IT DOES

| Function | Technical Action | Field Application |
|---|---|---|
| Source (mA) | Outputs a precise mA signal using internal battery power. | Simulates a transmitter to test wiring and PLC cards. |
| Simulate (mA) | Acts as a loop-powered transmitter (draws power from loop). | Tests the signal path with a 2-wire loop powered from the PLC. |
| Measure (mA) | Reads loop current (similar to DMM mA mode). | Verifies actual transmitter output. |
| Supply 24 VDC | Provides loop power to a device. | Bench-tests a transmitter without a panel connection. |
| Step / Ramp | Automatically increments through mA values. | Verifies full-span calibration and PLC scaling. |

## SOURCE VS. SIMULATE—THE CRITICAL DIFFERENCE

This is the most common confusion with loop calibrators.

- **Source mode:** the calibrator generates its own mA output using its internal battery. It does not need loop power from the PLC or power supply. Use source mode when you have disconnected the transmitter and need to inject a test signal into the loop independently.
- **Simulate mode:** the calibrator behaves like a 2-wire loop-powered transmitter. It draws its operating power from the loop supply (the PLC card or external power supply) and modulates that current to the value you set. Use simulate mode when you want to replace the transmitter in a live loop without providing separate power.

**⚠ WARNING:** If you connect in source mode to a live loop, you now have two current sources fighting each other. Results are unpredictable and potentially damaging to the card. Know which mode you're in.

### THE SUBSTITUTION TEST WORKFLOW

When a transmitter is suspected to be faulty, follow this systematic isolation process:

1. Disconnect transmitter from loop terminals. This isolates the transmitter from the signal path.
2. Connect calibrator in simulate mode. The calibrator now acts as the transmitter.
3. Source 4.0 mA (0%). PLC should show 0%—confirms wiring and card are good.
4. Source 12.0 mA (50%). PLC should show 50%.
5. Source 20.0 mA (100%). PLC should show 100%.

6. If PLC tracks correctly at all three points, the signal path is healthy and the transmitter has failed.
7. If PLC does not track, the fault is in wiring or input card, not the transmitter.

**The Verdict:**

• If the PLC tracks correctly at all points, the **transmitter has failed**.

• If the PLC does not track, the fault lies in the **wiring or the input card**.

## SPAN VERIFICATION (0-100 PSI Example)

Use the step function to verify the scaling across the entire range.

| mA Output | % of Span | Expected HMI Value |
|---|---|---|
| 4.0 mA | 0% | 0 PSI |
| 8.0 mA | 25% | 25 PSI |
| 12.0 mA | 50% | 50 PSI |
| 16.0 mA | 75% | 75 PSI |
| 20.0 mA | 100% | 100 PSI |

If the HMI value tracks correctly across all five points, the signal path and scaling are healthy.

💡 **PRO TIP:** Bench-test any 2-wire transmitter using a calibrator's **24 VDC power supply** before installation. Verify it is functional before climbing to the installation point.

## TOOL 3: HART COMMUNICATOR

A HART communicator enables a digital "conversation" with a smart transmitter over the existing 4-20 mA signal wires. This allows for configuration and diagnostics without additional wiring, process interruption, or disconnecting the loop.

See Chapter 13 (Analog Loops) for a full explanation of the HART protocol. This section covers the tool itself.

### CONNECTING THE COMMUNICATOR

To establish a connection, attach the communicator leads across the transmitter terminals, the loop resistor, or any accessible terminal block within the loop.

- **Polarity:** You must observe the +/− **markings** for a successful connection.
- **Impedance Requirement:** The loop requires a minimum of approximately **250 Ω** for communication. If the PLC analog input card does not provide sufficient impedance, add a 250 Ω resistor in series temporarily.

⚠ SAFETY SYSTEM WARNING: Do not connect to a live loop integrated with a safety system without verifying the system is bypassed or in maintenance mode. While the communication signal is generally harmless, always consult safety documentation first.

## WHAT YOU CAN READ WITH A HART COMMUNICATOR

| Parameter | Menu Location | Diagnostic Value |
|---|---|---|
| Primary Variable (PV) | Device Variables | Measured value; compare directly to the HMI. |
| Output Current (mA) | Device Variables | The actual mA output the transmitter is generating. |
| Device Status | Diagnostics | Reports internal faults and sensor degradation flags. |
| Range (LRV/URV) | Basic Setup | Confirms configured span to catch range mismatches. |
| Sensor Type | Basic Setup | Identifies the sensor (e.g., Pt100, Type K) to catch config errors. |
| Damping Setting | Basic Setup | High values can cause sluggish process response. |
| Engineering Units | Basic Setup | Confirms exactly what the mA signal represents. |
| Loop Test | Diagnostics / Test | Forces a specific mA output to test the signal path. |

## THE THREE-MINUTE DIAGNOSTIC SEQUENCE

When troubleshooting a suspected transmitter issue, follow this rapid sequence to isolate the fault:

1. **Connect** the HART communicator.
2. **Read Device Status** to check for self-reported internal faults.
3. **Compare PV to HMI:** If they match, the fault is likely downstream of the transmitter.
4. **Compare Output Current to Meter:** If the HART reading and your physical meter reading differ, the transmitter has an output stage problem.
5. **Confirm Configuration:** Verify that the **Range** and **Sensor Type** match the actual application.

💡 Pro Tip: **PRO TIP:** Use the **Loop Test** function to force the transmitter to 4 mA, 12 mA, or 20 mA. This performs a signal path substitution test without needing a separate loop calibrator.

## TOOL 4: CLAMP-ON mA METER

A clamp-on mA meter measures 4-20 mA loop current without breaking the circuit. By clipping the jaws around a single conductor, you can read the current via the magnetic field generated by the flow.

WHEN TO REACH FOR IT

| Scenario | Technical Advantage |
|---|---|
| Running Process | No disconnection required; keeps the process online. |
| Difficult Access | Measure at any accessible point along the cable. |
| High-Volume Checks | Faster than opening multiple terminals for spot-checks. |
| Output Verification | Non-invasive verification of transmitter performance. |
| M12 D-coded | For Industrial Ethernet/IO-Link repairs. |
| Panel Receptacles | To replace damaged bulkheads on junction boxes. |

## USING IT CORRECTLY

The accuracy of this tool depends entirely on proper lead isolation.

- **Single Conductor Rule:** You must clamp around **one wire only**. If you clamp around both conductors of a 2-wire loop, the magnetic fields cancel out, resulting in a reading of zero.
- **Shielded Cable:** You must clamp around the internal signal conductors, **not the outer shield**.
- **Accuracy Limits:** Most industrial models are accurate to **±0.1 mA** within the 4-20 mA range. Accuracy often degrades significantly below 1 mA.

Pro Tip: Label the conductors at every junction box you work in. A clamp-on meter is useless if you can't identify which wire is the signal conductor.

## TOOL 5: INSULATION RESISTANCE TESTER (MEGGER)

A "Megger" applies high DC voltage (500–1000 VDC) to stress insulation and measure resistance in **Megohms (M$\Omega$)**. While a standard DMM checks for shorts, the Megger reveals degradation that hasn't fully failed yet.

Your standard DMM measures resistance in ohms or kilohms. Insulation resistance is in megohms—millions of ohms. A DMM cannot perform this test because it doesn't apply enough voltage to stress the insulation. The megger applies enough voltage to reveal degradation that would not show on a standard ohmmeter.

## WHAT GOOD INSULATION LOOKS LIKE

| Insulation Condition | Resistance Reading | Interpretation |
|---|---|---|
| Excellent | > 1,000 MΩ | New cable, dry conditions. |
| Healthy | 100–1,000 MΩ | Good cable, normal service. |
| Marginal | 10–100 MΩ | Acceptable, monitor closely. |
| Action Required | 1–10 MΩ | Degraded—plan replacement. |
| Failing | < 1 MΩ | Replace immediately; do not return to service. |
| Load Cell Spec | > 5,000 MΩ | Special standard for precision sensors. |

## WHEN TO USE IT

- Instrument cable after flood or water ingress—test insulation integrity after moisture exposure. Test at 500 VDC.
- Sensor cable after suspected damage—test conductor-to-ground insulation. Test at 500 VDC.
- Load cell cable (see Chapter 18)—test insulation from strain gauges to cable shield. Test at 500 VDC.
- Motor winding integrity—test winding-to-frame insulation. Test at 500–1000 VDC.
- New cable installation verification—confirm no installation damage. Test at 500 VDC.

## ⚠ CRITICAL SAFETY RULES

- **Isolate Equipment:** ALWAYS disconnect the device (PLC card, transmitter) before testing. 500V will destroy electronics.
- **Discharge:** Cables hold a capacitive charge after testing. Short to ground through a 10 kΩ resistor before touching.
- **IS Barriers:** Never Megger cables connected to Zener or intrinsically safe barriers.

⚡ Shop-Floor Wisdom: A cable that passes a DMM continuity check and fails a megger test has degraded insulation that hasn't shorted yet. The continuity check gives it a passing grade. The megger tells the truth.

## TOOL 6: LAPTOP WITH PLC SOFTWARE

The laptop is the ultimate diagnostic bridge. It allows you to see the "mental state" of the controller—what it knows, what it's doing, and why it's stopped.

### MINIMUM REQUIREMENTS

| Requirement | Technical Specification |
|---|---|
| Programming Software | **Rockwell**: Studio 5000, RSLogix 5000/500. <br> **Siemens**: TIA Portal, STEP 7. <br> **Others**: Mitsubishi GX Works, Beckhoff TwinCAT. |
| Communication Drivers | **Rockwell**: RSLinx Classic or Enterprise. <br> **Siemens**: Integrated within TIA Portal. |
| Network Adapter | Standard Built-in Ethernet (RJ45) for modern PLCs. |
| Serial Connectivity | USB-to-Serial Adapter: Necessary for legacy protocols like DF1, RS-232, or proprietary serial links. |
| Authorization | Verified software license or key for "Edit" mode; confirm plant-specific requirements before deployment. |
| Panel Receptacles | To replace damaged bulkheads on junction boxes. |

## GETTING ONLINE

- **Subnet Match:** Ensure your laptop IP (e.g., 192.168.1.50) is on the same subnet as the PLC.
- **Ping Test:** Verify connectivity via command prompt. No ping = no PLC access.
- **Browse:** Use drivers (RSLinx, TIA Portal) to locate the processor.
- **Monitor Mode:** Always default to **Monitor Mode** to observe logic without risk of accidental changes.
- **Backup:** Upload and save the current program with a timestamp before doing anything else.

⚠️ Going online in Program mode gives you the ability to make changes that take effect immediately on the running machine. One accidental keystroke can change a setpoint, delete a rung, or force an output. Default to Monitor mode. Switch to Program mode only when required, and only when you are certain of what you are about to change.

## WHAT TO DO ONLINE

Once you have established a connection to the PLC, utilize these specific functions to diagnose the system state.

| Function | Method | Diagnostic Value |
|---|---|---|
| Watch Tag/Bit | Select tag in the Watch window. | Monitors real-time state of I/O points or internal variables. |
| Monitor Rung Logic | Online ladder diagram view. | Identifies where logic is blocked by showing true/false conditions. |
| Check Force Table | Controller Properties → Tasks → Force table. | Reveals if any bits are being manually forced on or off. |
| Diagnostic Buffer | Controller Properties → Diagnostics (Rockwell/Siemens). | Provides a historical log of system faults with timestamps. |
| Check Scan Time | Controller Properties → Tasks. | Confirms the PLC scan time is within a healthy operational range. |
| Cross-Reference | Right-click tag → Cross Reference. | Lists every location in the program that reads or writes that tag. |

## Strategy: Going Online on an Unfamiliar Machine

Avoid the common mistake of scrolling aimlessly through ladder rungs. Follow this high-efficiency workflow instead:

- **Read the Diagnostic Buffer First:** Let the machine tell you what it already knows about the fault.
- **Monitor the Step Variable:** Watch the sequence state live; find exactly where the program stops advancing.
- **Cross-Reference the Block:** Find every location where the blocking tag is written to identify what is holding it false.
- **Verify the Force Table:** Ensure an "impossible" machine behavior isn't simply the result of a forced bit before assuming hardware failure.

Pro Tip: The cross-reference function is the most underused tool in PLC software. Any tag, any address—right-click and find everywhere it's used in the program. A search that would take an hour scrolling takes 10 seconds with cross-reference.

*APPENDIX H: The Technician's Field Kit* 299

# FIELD KIT SUMMARY

| Tool | Minimum Specifications | Useage |
|---|---|---|
| Multimeter | True RMS, CAT III minimum. | Used on every call. |
| Loop Calibrator | Source + Simulate + 24V supply. | Any analog loop fault. |
| HART Communicator | Universal or DD-based. | Smart transmitter diagnostics. |
| Clamp-On mA Meter | ±0.1 mA resolution (4-20 mA). | Live loop checks. |
| Megger | 500 VDC and 1000 VDC test voltage. | Cable and motor integrity. |
| Laptop | Current, licensed PLC software. | Any logic or I/O fault. |

## Shop-Floor Essentials: The Cart

These items are not test instruments, but they are critical for rapid field repairs:

- **Mechanical Tools:** Small flathead and Phillips screwdrivers, Torx driver set, and needle-nose pliers.
- **Wiring:** Wire stripper/crimper, cable ties, and label tape.
- **Visibility & ID:** Flashlight/headlamp and permanent markers for identifying cables.
- **Documentation:** A pocket notebook to record findings and changes.
- **Spares:** Assorted fuses and M12 cordsets (3-pin and 4-pin).

⚡ Shop-Floor Wisdom: The notebook is not optional. Memory fails and shifts change—the next technician on the machine relies entirely on the notes you leave behind.

# APPENDIX I: ISA INSTRUMENT TAG QUICK REFERENCE

Every instrument in an industrial plant has a tag—a short alphanumeric code that identifies what it measures and what it does. Tags appear on P&IDs, loop sheets, instrument indexes, PLC tag databases, and HMI screens. If you can read a tag, you can find the instrument, understand its function, and locate it in the program without asking anyone.

Tag structure follows ISA 5.1, the dominant standard in North American and international process and discrete manufacturing. Once you know the letter code system, any tag in any plant is readable.

The pages that follow will provide an example and overview of ISA 5.1 Instrument Tagging Logic.

## TAG STRUCTURE

A tag is built from two parts: a letter combination followed by a loop number.

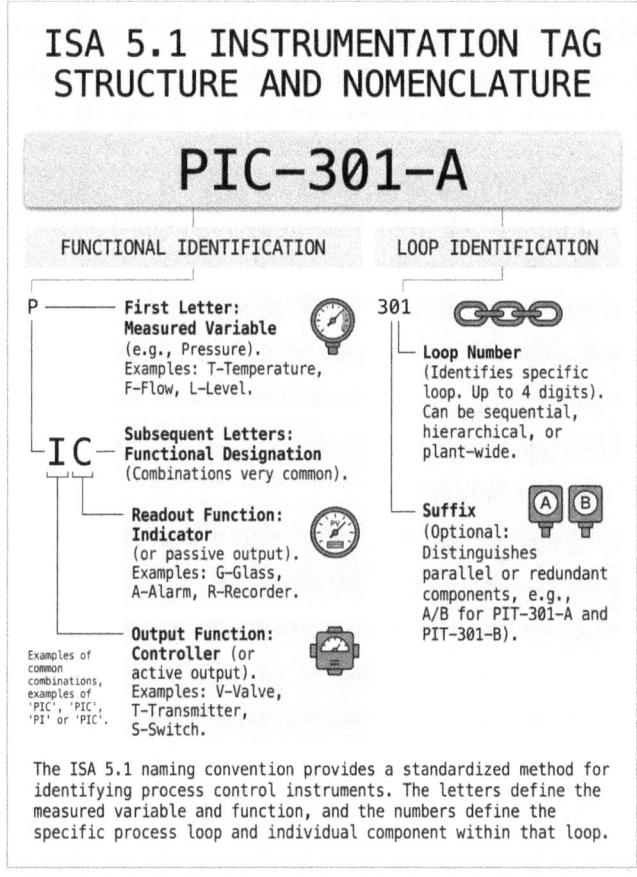

The loop number may include an area prefix (e.g., FT-2-101 = Flow Transmitter, Area 2, Loop 101). Some plants use four-digit loop numbers; others use three. The letter code is always the same regardless of plant numbering convention.

# APPENDIX I: ISA Instrument Tag Quick Reference

## PROFESSIONAL ISA 5.1 INSTRUMENT TAGGING LOGIC AND FIELD REFERENCE

**FIRST LETTER = MEASURED VARIABLE | ALL SUCCEEDING LETTERS = FUNCTIONS PERFORMED ON THAT VARIABLE**

### 1. FIRST LETTER—MEASURED VARIABLE

| LETTER | MEASURED VARIABLE AND COMMON EXAMPLES |
|---|---|
| A | Analysis (pH, $O_2$, conductivity, $CO_2$) |
| B | Burner / Combustion |
| C | Conductivity (user's choice) |
| D | Density / Specific Gravity (user's choice) |
| E | Voltage (EMF) |
| F | Flow Rate |
| G | Gauging (user's choice) |
| H | Hand (manual) |
| I | Current (Electrical) |
| J | Power |
| K | Time / Time Schedule |
| L | Level |
| M | Moisture / Humidity (user's choice) |
| N | User's Choice |
| O | User's Choice |
| P | Pressure / Vacuum |
| Q | Quantity / Totalizer |
| R | Radiation |
| S | Speed / Frequency |
| T | Temperature |
| U | Multivariable |
| V | Vibration / Mechanical Analysis |
| W | Weight / Force |
| X | Unclassified |
| Y | Event / State / Presence |
| Z | Position / Dimension |

### 2. SUCCEEDING LETTERS—FUNCTION MODIFIERS

| LETTER | FUNCTION PERFORMED AND DESCRIPTION |
|---|---|
| A | Alarm—Generates an alarm output. |
| C | Control—Controls the process variable. |
| D | Differential—Measures difference between two points. |
| E | Element (Primary)—The sensing element itself (no signal output). |
| F | Ratio / Fraction—Ratio control or ratio measurement. |
| G | Glass / Gauge (Viewing)—Local sight glass or visual indicator. |
| H | High—High setpoint or high state. |
| I | Indicate—Local or remote indication (display). |
| N | Scan—Multipoint scanning. |
| O | Orifice / Restriction—Flow restriction element. |
| P | Point (Test Connection)—Test or sampling connection. |
| Q | Integrate / Totalize—Accumulates over time. |
| R | Record—Data recording (chart, historian). |
| Z | Driver / Actuator—Final control element driver. |

### 3. ALARM AND LIMIT MODIFIERS

| MODIFIER | FUNCTION | DESCRIPTION |
|---|---|---|
| H | High | First high alarm or high switch setpoint. |
| HH | High High | Second high alarm, often a hardwired trip. |
| L | Low | First low alarm or low switch setpoint. |
| D | Differential | Difference between two measurements. |
| R | Rising | Triggered on rising signal. |
| F | Falling | Triggered on falling signal. |

### 5. READING A LOOP NUMBER

Loop numbers are plant-specific but follow common conventions:
- FT-101 = Flow Transmitter, loop 101 (simple sequential)
- FT-2-101 = Flow Transmitter, Area 2, loop 101 (area-loop)
- TT-U3-045 = Temperature Transmitter, Unit 3, loop 045 (unit-loop)
- PT-1201 = Pressure Transmitter, sheet 12, loop 01 (P&ID sheet-loop)

When you encounter an unfamiliar numbering convention, find the instrument index (a spreadsheet or document that lists every tag, its location, its range, and its P&ID reference). Every plant that follows good documentation practice has one.

### 4. COMMON TAG COMBINATIONS—FIELD REFERENCE

| TAG | TAG | DESCRIPTION |
|---|---|---|
| **Flow** | | |
| FE | Flow Element | Primary element only (orifice plate, pitot tube)—no signal output. |
| FT | Flow Transmitter | Converts flow measurement to 4-20 mA signal. |
| FI | Flow Indicator | Local display of flow rate. |
| FIC | Flow Indicating Controller | Displays flow and controls it (PID loop). |
| FS | Flow Switch | Switches output at a setpoint. |
| FSH / FSL | Flow Switch High / Low | Switches when flow exceeds or drops below setpoint. |
| FQ | Flow Totalizer | Accumulates total flow over time. |
| FCV | Flow Control Valve | Control valve for flow regulation. |
| FAH / FAL | Flow Alarm High / Low | |
| **Temperature** | | |
| TE | Temperature Element | Thermocouple or RTD sensor only—no signal output. |
| TT | Temperature Transmitter | Converts temperature to 4-20 mA signal. |
| TI | Temperature Indicator | Local thermometer or display. |
| TIC | Temperature Indicating Controller | Displays temperature and controls it (PID). |
| TS | Temperature Switch | Switches at a temperature setpoint. |
| TSH / TSL | Temperature Switch High / Low | Switches when flow exceeds or drops leet. |
| TSHH | Temperature Switch High High | Emergency high temperature trip. |
| TW | Thermowell | Protective well for temperature sensor |
| TAH / TAL | Temperature Control Valve | Temperature Alarm High / Low |
| **Level** | | |
| LT | Level Transmitter | Converts level to 4-20 mA signal. |
| LI | Level Indicator | Local level gauge or sight glass. |
| LG | Level Gauge (Glass) | Sight glass—visual only, no signal. |
| LIC | Level Indicating Controller | Displays level and controls it. |
| LS | Level Switch | Switches at a level setpoint. |
| LSH / LSL | Level Switch High / Low | Switches at a level setpoint. |
| LSHH / LSLL | Level Switch High High / Low Low | Emergency trips |
| LCV | Level Control Valve | Emergency trips |
| LAH / LAL | Level Alarm High / Low | Level Alarm High / Low |
| **Valves and Final Elements** | | |
| HV | Hand Valve | Manual, operator-actuated |
| CV | Control Valve | Automated control valve (generic) |
| SOV | Solenoid Valve | Solenoid-operated on/off valve |
| XV | On-Off Valve | Automated block valve, not throttling |
| FCV / PCV | TCV / LCV | Pressure / Temperature / Level Control Valve |
| PRV | Pressure Relief Valve | Safety/relief valve (mechanical) |
| **Discrete and Electrical** | | |
| ZT | Position Transmitter | Analog position feedback (4-20 mA) |
| ZS | Position Switch | Limit switch—discrete position feedback |
| ZSO / ZSC | Position Switch Open / Closed | Position Switch |
| HS | Hand Switch | Operator pushbutton or selector switch |
| SS | Speed Switch | Trips on speed setpoint |
| SSH / SSL | Speed Switch | Speed Switch High / Low |
| **Analysis** | | |
| AT | Analyzer Transmitter | Outputs analysis result as 4-20 mA |
| AIC | Analyzer Indicating Controller | Controls based on analysis value |
| AAH / AAL | Analyzer Alarm | Analyzer Alarm High / Low |
| **Weight and Force** | | |
| WT | Weight Transmitter | Load cell / scale transmitter output |
| WI | Weight Indicator | Local weight display |
| WIC | Weight Indicating Controller | Controls based on weight (batching) |
| WS | Weight Switch | Trips at a weight setpoint |
| WAH / WAL | Weight Alarm | Weight Alarm High / Low |

### 6. QUICK DECODE EXAMPLES

| TAG | DESCRIPTION |
|---|---|
| FT-101 | Flow Transmitter, Loop 101 |
| TIC-204 | Temperature Indicating Controller, Loop 204 |
| PSHH-312 | Pressure Switch High High, Loop 312—emergency trip |
| LT-055 | Level Transmitter, Loop 055 |
| PDT-118 | Differential Pressure Transmitter, Loop 118 |
| TE-401 | Temperature Element (bare sensor), Loop 401—no output, connects to TT or TIC |
| ZSC-22 | Position Switch Closed, Loop 22—cylinder or valve closed confirmation |
| XV-501 | On-Off Valve, Loop 501—automated block valve |
| FCV-088 | Flow Control Valve, Loop 088 |
| LSLL-007 | Level Switch Low Low, Loop 007—emergency pump protection |

## COMMON MISTAKES IN THE FIELD

| The Mistake | Consequence |
|---|---|
| Confusing TE and TT | A TE (Element) has no signal output; you cannot measure 4-20 mA at a thermowell. The TT (Transmitter) is where the loop signal lives. |
| Confusing FS and FT | An FS is a discrete switch (ON/OFF). An FT is an analog transmitter (4-20 mA). They look similar on P&IDs but require different tools. |
| Treating HH/LL as H/L | HH and LL are typically hardwired safety trips, not standard PLC alarms. Bypassing them requires distinct, strict procedures. |
| Ignoring Area Prefixes | FT-101 and FT-2-101 are often completely different instruments. Always verify against the Instrument Index. |
| Assuming ZS Technology | ZS means Position Switch (function). It could be a proximity sensor, limit switch, or reed switch—the tag identifies the role, not the hardware. |
| Beckhoff / Codesys | Online → Force Values table. |

💡 Pro Tip: When handed a work order with an unfamiliar tag, decode the letters first. Knowing the signal type (analog vs. discrete) tells you exactly which tools to pull from your kit before heading to the floor.

⚡ Shop-Floor Wisdom: The **Tag** is the address. The **P&ID** is the map. The **Loop Sheet** is the wiring diagram. With these three pieces of documentation, you can troubleshoot an instrument in any plant on earth, even if you've never set foot in the facility before.

# APPENDIX J: COMMON PLC ERROR CODES

When a PLC faults, the machine stops and the pressure's on. The fault code is your starting point—but only if you know what it means and what to check first.

This appendix covers the most common error codes you'll encounter on the three major PLC platforms: Rockwell/Allen-Bradley, Siemens, and Schneider/Modicon. These aren't comprehensive fault manuals—those exist in the manufacturer documentation. This is a field reference for the faults that cause 80% of your downtime. These error codes are accurate as of publication (2026), but PLC firmware updates and new controller models can change diagnostic behavior. Always cross-reference with the manufacturer's current documentation for your specific controller model and firmware revision.

**What's included:**

**Rockwell/Allen-Bradley** (pages 302-303): Common major and minor faults for ControlLogix, CompactLogix, and MicroLogix. Covers instruction errors, I/O configuration mismatches, watchdog timeouts, and communication faults.

**Siemens S7** (pages 304-305): Diagnostic events for S7-300, S7-400, S7-1200, and S7-1500 series. Covers hardware interrupts, cycle time exceeded, PROFIBUS/PROFINET errors, and system faults.

**Schneider/Modicon** (page 306): System faults for M340, M580, and

Quantum controllers. Less detailed diagnostics than Rockwell or Siemens, but covers the categories you'll encounter on OEM equipment.

# ROCKWELL AUTOMATION / ALLEN-BRADLEY

## Common Fault Codes

These are the most common major and minor fault codes you'll encounter on Allen-Bradley PLCs (ControlLogix, CompactLogix, MicroLogix). When a major fault occurs, the PLC stops and the OK light flashes red. Minor faults log but don't halt execution.

| Code | Description | Common Cause | First Action |
|---|---|---|---|
| 0001 | Instruction Execution Error | Math overflow, divide by zero, array bounds | Check logic around fault instruction; verify tag values |
| 0002 | RAM Parity Error | Memory corruption, failing hardware | Power cycle PLC; if repeats, replace processor |
| 0003 | I/O Module Configuration Mismatch | Module replaced with different type | Match module type in config or update I/O tree |
| 0004 | Watchdog Timeout | Logic scan taking too long | Reduce scan time; check for infinite loops |
| 0010 | Battery Low | Backup battery needs replacement | Replace battery; verify battery connection |
| 0020 | Loss of I/O Communication | Module pulled, failed, or unpowered | Check module OK light; verify power; reseat module |
| 0031 | Motion Axis Fault | Servo drive fault, wiring issue | Check drive faults; verify feedback wiring |

APPENDIX J: Common PLC Error Codes    307

## ROCKWELL COMMON ERROR CODES CONTINUED

| Code | Description | Common Cause | First Action |
|---|---|---|---|
| 0042 | Safety Signature Mismatch | Safety program changed without validation | Re-validate safety signature in software |
| 0054 | Motion Planner Fault | Coordinated motion error | Check motion coordinate system configuration |
| 0080 | Program File Missing | Main routine or subroutine not found | Verify program structure; restore missing routine |
| 0090 | Memory Module Error | CompactFlash or SD card issue | Reseat card; replace if failed; verify format |
| 1756-xx | Module-Specific Fault | Varies by module | Check module diagnostics in software |
| ENet Timeout | EtherNet/IP Timeout | Network overload, cable issue, switch problem | Check RPI settings; verify network cable; ping device |

Rockwell Pro Tip: Most major faults halt the PLC immediately. Always check the fault log in Studio 5000 or RSLogix—it shows the exact instruction and rung where the fault occurred. This is your starting point.

Minor Faults: Minor faults don't stop the PLC but indicate something is wrong. Common examples: controller battery low, I/O connection faulted (module still working but connection lost), or a single instruction failing while the rest of the program runs.

# SIEMENS S7 SERIES

## Common Diagnostic Events

Siemens uses diagnostic events and system faults rather than numbered fault codes. Events appear in the diagnostics buffer and can trigger STOP mode. S7-300, S7-400, S7-1200, and S7-1500 use similar diagnostic systems accessed through TIA Portal.

| Event ID | Description | Common Cause | First Action |
|---|---|---|---|
| 16#4xxx | Hardware Interrupt | Module failure, I/O error | Check module LED; verify wiring; check diagnostics |
| 16#6xxx | Cycle Time Exceeded | Scan time too long | Increase watchdog time or optimize program |
| 16#7xxx | Communication Error | PROFIBUS/ PROFINET fault | Check network cable; verify device address |
| 16#8xxx | Programming Error | Logic error, array access | Check program at faulted block; verify tag ranges |
| 16#A0xx | Battery Low | Backup battery needs replacement | Replace battery; verify connection |
| 16#C4xx | PROFINET Device Missing | Device offline or unpowered | Verify device power; check network cable |

APPENDIX J: *Common PLC Error Codes* 309

| Event ID | Description | Common Cause | First Action |
|---|---|---|---|
| 16#D2xx | Module Configuration Error | Wrong module type or parameters | Verify module matches HW config |
| 16#D4xx | PROFIBUS Master Failure | DP master offline | Check master power; verify bus termination |
| Memory Reset | Total Memory Loss | Battery dead during power loss | Download program; replace battery |
| OB Error | Organization Block Missing | Required OB not in program | Add missing OB or disable function |
| Rack/ Slot Fault | Module Not Responding | Module failure or loose connection | Reseat module; check backplane |
| Safety Fault | F-Device Discrepancy | Safety validation mismatch | Re-validate safety program |

Siemens Pro Tip: Diagnostic events are stored in the diagnostics buffer in TIA Portal. Unlike Rockwell which stops immediately on major faults, Siemens PLCs can be configured to handle certain errors without stopping—they call error-handling Organization Blocks (OBs) instead. If an OB isn't present for a specific error, the PLC goes to STOP.

LED Indicators: Siemens CPUs have LED status lights. SF (System Fault) = red means error in diagnostics buffer. BF (Bus Fault) = red means PROFIBUS/PROFINET problem. DC5V = green means power OK. RUN = green means executing, STOP = yellow means halted.

# SCHNEIDER ELECTRIC / MODICON

## Common System Faults

Schneider/Modicon PLCs (M340, M580, Quantum) use fault codes displayed on the controller or in Unity Pro/Control Expert software. Less common than Rockwell or Siemens in many plants, but you'll encounter them on OEM equipment and certain industries.

| Fault | Description | Common Cause | First Action |
|---|---|---|---|
| APP STOPPED | Application Halted | User stopped or fault condition | Check fault log; verify stop command |
| ERR | I/O Error | Module fault or configuration mismatch | Check module status LEDs; verify config |
| BAT | Battery Fault | Low or missing battery | Replace battery; verify connection |
| COMM | Communication Fault | Network timeout or device offline | Check Modbus/ Ethernet cable; ping device |
| CONF | Configuration Mismatch | Hardware does not match config | Match physical modules to Unity config |
| SYS | System Fault | CPU error or watchdog timeout | Power cycle; check scan time; review diagnostics |

Schneider Pro Tip: Modicon controllers often appear on packaged equipment (OEM skids, conveyors, specialized machinery). The diagnostic information is less detailed than Rockwell or Siemens, so you'll spend more time isolating the fault manually. The ERR light is your starting point—it means something in the I/O system failed.

LED Indicators: RUN = green (executing), ERR = red (I/O or config fault), BAT = amber (battery low), I/O = green (modules OK). Unlike Rockwell and Siemens which give you detailed fault codes, Schneider often just gives you a category (ERR, SYS, COMM) and you troubleshoot from there.

# GLOSSARY OF TERMS

**2-Wire Transmitter**—A smart transmitter that uses the same two wires for both power and signal. The device draws its operating current (typically 3.6-4 mA) from the loop and modulates the current to send the process measurement (4-20 mA). The most common transmitter type in industrial instrumentation.

**3-Wire RTD**—An RTD wiring configuration that uses three wires to compensate for lead resistance. Two wires connect to one side of the RTD element, one wire connects to the other side. The PLC card measures lead resistance and subtracts it from the reading. Standard for industrial temperature measurement where wire runs exceed 10 feet.

**4-Wire RTD**—The most accurate RTD wiring configuration. Uses separate wire pairs for excitation and measurement, completely eliminating lead resistance errors. Required for precision temperature measurement or very long wire runs. More expensive due to additional wiring and dedicated input channels.

**4-20 mA**—The standard current loop signal in industrial instrumentation. 4 mA represents zero (e.g., 0 PSI), 20 mA represents full scale (e.g., 100 PSI). The reason it starts at 4 instead of 0: a reading of 0 mA always means something is broken, so you can distinguish "zero flow" from "dead transmitter."

**A/D Converter (Analog-to-Digital Converter)**—The circuit inside an analog input card that converts a continuous electrical signal into a digital number (raw counts) the PLC can process.

**Absolute Pressure (PSIA)**—Pressure measured relative to a perfect vacuum. Atmospheric pressure at sea level is approximately 14.7 PSIA. Used in vacuum systems and altitude-sensitive applications.

**AC (Alternating Current)**—Electrical current that reverses direction periodically. Standard in power distribution (120 VAC, 480 VAC). Some older discrete I/O systems use 120 VAC inputs instead of 24 VDC.

**AI (Analog Input)**—A PLC input channel that reads a proportional signal (typically 4-20 mA or 0-10 VDC) and converts it to raw counts for processing.

**Analog Loop**—The complete signal path from field transmitter through wiring, terminal blocks, and input card to the PLC processor. When troubleshooting, verify each segment of the loop systematically: power at transmitter, current in the field, raw counts at the card.

**AO (Analog Output)**—A PLC output channel that writes a proportional signal (typically 4-20 mA) to drive control valves, VFDs, or other proportional actuators.

**Capacitive Sensor**—A proximity sensor that detects changes in an electrostatic field. Unlike inductive sensors, capacitive sensors can detect non-metallic targets: plastics, liquids, powders, glass, and wood. Requires sensitivity adjustment for each target material.

**CAT Rating**—The safety category rating on a multimeter (CAT I through CAT IV) that indicates the level of transient voltage protection. Higher CAT numbers mean the meter can safely handle larger voltage spikes. Never use a CAT II meter on industrial power circuits.

**Coriolis Meter**—A flow meter that measures mass flow directly by vibrating a tube and detecting the twist caused by fluid flowing through it. Highly accurate, handles nearly any fluid, but expensive and sensitive to entrained gas.

**Custody Transfer**—A flow measurement point where product ownership changes hands (e.g., between a pipeline supplier and a buyer).

Requires the highest measurement accuracy and is usually subject to regulatory standards.

**D/A Converter (Digital-to-Analog Converter)**—The circuit inside an analog output card that converts a digital value from the PLC into a proportional electrical signal (typically 4-20 mA).

**DC (Direct Current)**—Electrical current that flows in one direction. Standard for industrial control circuits (24 VDC) and sensor power.

**DCS (Distributed Control System)**—A process control architecture common in refineries, chemical plants, and power generation. Similar to a PLC in function but designed for continuous process control rather than discrete manufacturing.

**DI (Discrete Input)**—A PLC input that reads an ON/OFF state from a field device such as a switch, sensor, or relay contact.

**Diagnostic Buffer**—A timestamped fault log maintained by the PLC processor. Records mode changes, hardware faults, and system events. Your first stop for intermittent or recurring problems—read it before touching the logic.

**Diaphragm Seal**—A flexible membrane that isolates a pressure transmitter from corrosive, viscous, or high-temperature process fluids. Fill fluid (typically silicone oil) transmits pressure from the diaphragm to the sensor element.

**Dielectric Constant**—A measure of how strongly a material affects an electrostatic field. Water has a high dielectric constant (~80), making it easy for capacitive sensors to detect. Dry powders and plastics have low dielectric constants and are harder to detect.

**Differential Pressure (PSID)**—The difference between two pressure measurements. Used for flow measurement (across an orifice plate), filter monitoring (across a filter element), and level measurement in pressurized vessels.

**DIN Rail**—A standardized metal rail used to mount industrial components (relays, terminal blocks, transmitters, power supplies) inside control panels. Named after the German Institute for Standardization.

**DO (Discrete Output)**—A PLC output that switches a field device ON or OFF. Can be relay, transistor, or triac type, each with different load ratings and failure modes.

**Emissivity**—A surface property (0 to 1.0) that describes how efficiently a material radiates infrared energy. Critical for infrared temperature sensors—a polished metal surface (emissivity ~0.15) will read far lower than actual temperature if the sensor is calibrated for a matte surface (emissivity ~0.95).

**Encoder**—A device that converts mechanical position or motion into electrical signals. Used for position tracking, speed measurement, and pulse counting. Requires dedicated high-speed input channels on the PLC.

**Flush Mount**—A proximity sensor installation where the sensing face sits flush with the surrounding metal. Reduces sensing range but protects the sensor from physical damage. Requires a flush-rated sensor.

**Force Table**—A PLC feature that overrides input or output states regardless of program logic. Essential for testing but dangerous if forgotten—a forced output can keep a device energized (or de-energized) indefinitely. Always check the force table first when troubleshooting unexplained behavior.

**Function Block Diagram (FBD)**—A graphical PLC programming language that uses interconnected function blocks (AND, OR, timers, counters, PID) to build control logic. Common in process control and Siemens platforms.

**Gauge Pressure (PSIG)**—Pressure measured relative to atmospheric pressure. A tire at 30 PSIG is 30 PSI above atmosphere. Most industrial pressure sensors measure gauge pressure.

## Glossary of Terms

**Ground Loop**—An unwanted current path created when a signal shield or circuit is grounded at more than one point. Causes noise, erratic readings, and measurement errors in analog signals. Fix: ground the shield at one end only.

**HART (Highway Addressable Remote Transducer)**—A digital communication protocol that rides on top of the standard 4-20 mA signal without disturbing it. Lets you read transmitter diagnostics, configuration, and process values through the same two wires carrying the analog signal.

**HMI (Human-Machine Interface)**—The touchscreen or panel display that operators use to monitor and control a process. Displays values calculated from PLC data—if the HMI shows the wrong number but raw counts are correct, the problem is in scaling or display configuration, not the field device.

**Hysteresis**—The intentional dead band in a switching device that prevents rapid on/off cycling at the switching threshold. Example: A proximity sensor with 2mm hysteresis triggers at 8mm distance but won't release until the target moves to 10mm. Prevents chattering when a target vibrates at the sensing edge.

**I/O (Input/Output)**—The physical connection points between the PLC and the real world. Inputs read sensors and switches; outputs drive motors, valves, and indicators.

**Inductive Sensor**—A proximity sensor that detects metallic targets by generating a high-frequency electromagnetic field. The most common sensor type in discrete manufacturing. Does not require physical contact or sensitivity adjustment.

**Interposing Relay**—A relay placed between a PLC output and a high-power load. Lets a low-current PLC output (e.g., 0.5A transistor) control a high-current device (e.g., 10A contactor). Also provides electrical isolation between control and power circuits.

**IP Rating (Ingress Protection)**—A two-digit code indicating a device's protection against solids and liquids. IP67 means dust-tight and submersible to 1 meter. IP68 means dust-tight and submersible beyond 1 meter (depth specified by manufacturer).

**Ladder Logic**—The most common PLC programming language. Uses a graphical format resembling electrical relay diagrams with rungs, contacts, and coils. If you can read a relay schematic, you can read ladder logic.

**Live Zero**—The 4 mA baseline of a 4-20 mA signal that represents 0% of span. If you measure 4 mA, the process is at zero—transmitter working, wiring intact, power good. If you measure 0 mA, something is broken. Live zero lets you distinguish "process at zero" from "dead transmitter."

**Load Cell**—A transducer that converts mechanical force (weight) into an electrical signal using strain gauges arranged in a Wheatstone bridge. Common bridge resistances are 350 ohm and 700 ohm.

**Loop Calibrator**—A handheld instrument that can source, simulate, and measure 4-20 mA signals. Your second most important tool after a multimeter for analog loop troubleshooting. Can substitute for a transmitter to isolate wiring and card problems.

**MOV (Metal Oxide Varistor)**—A voltage-clamping device used to suppress transient voltage spikes across inductive loads (solenoids, contactors, relays). Protects output cards from back-EMF damage when the load switches off.

**NAMUR (Namur Open Architecture)**—A European instrumentation standard (NE 43) that defines extended current ranges for fault detection in 4-20 mA loops. NAMUR Low Fail (~3.6-3.8 mA) indicates sensor under-range or failure. NAMUR High Fail (~20.8-21 mA) indicates sensor over-range or failure. Modern transmitters and input cards support NAMUR ranges to provide diagnostic information beyond the normal 4-20 mA signal.

**NC (Normally Closed)**—A contact or switch that is closed (conducting) in its resting state and opens when actuated. An NC proximity sensor outputs a signal continuously and drops it when a target is detected.

**NO (Normally Open)**—A contact or switch that is open (not conducting) in its resting state and closes when actuated. An NO proximity sensor outputs no signal until a target is detected.

**NPN (Negative-Positive-Negative)**—A sensor output type where the output transistor connects the signal wire to 0V (ground) when active. The signal goes LOW when the sensor detects. Common in Asia and older equipment. Think: "Negative when active."

**NTC (Negative Temperature Coefficient)**—A type of thermistor whose resistance decreases as temperature increases. Common in HVAC equipment and refrigeration controls. Not interchangeable with RTDs or thermocouples.

**OEM (Original Equipment Manufacturer)**—The company that built the machine. OEM documentation is your primary reference for safety circuits, wiring diagrams, and spare parts.

**Orifice Plate**—A metal plate with a precisely sized hole inserted into a pipe to create a pressure drop proportional to flow. Flow is calculated from the differential pressure across the plate. Simple, no moving parts, but subject to wear and buildup.

**Photoeye (Photoelectric Sensor)**—A sensor that uses light (visible, infrared, or laser) to detect the presence, absence, or position of objects. Operating modes include through-beam, retroreflective, and diffuse.

**Piezoelectric Sensor**—A pressure sensor that responds to changes in pressure, not static pressure. The output naturally decays back to zero even if pressure remains constant. If a sensor "works for a moment then drifts back to zero," that is normal piezo behavior, not a failure.

**PLC (Programmable Logic Controller)**—The industrial computer that runs the machine's control logic. Reads inputs, executes the program, and writes outputs in a continuous scan cycle. The brain of the operation.

**PNP (Positive-Negative-Positive)**—A sensor output type where the output transistor connects the signal wire to +24V when active. The signal goes HIGH when the sensor detects. Most common in North America and Europe. Think: "Positive when active."

**Process Variable (PV)**—The physical quantity being measured or controlled: temperature, pressure, flow, level, speed, position, weight, etc.

**PSIA**—See Absolute Pressure.

**PSID**—See Differential Pressure.

**PSIG**—See Gauge Pressure.

**Raw Counts**—The integer value produced by an analog input card's A/D converter. This is the only value the PLC truly "knows." Everything displayed on the HMI is calculated from raw counts using a scaling formula. When readings are wrong, check raw counts first—if they track your measured current, the wiring and card are fine and the problem is in scaling.

**Registration Mark**—A printed mark on packaging material detected by a photoelectric sensor to trigger a cut, fold, or seal at the correct position. Requires high-speed, precise sensor response.

**Reynolds Number**—A dimensionless number that characterizes flow regime (laminar vs. turbulent). Relevant for vortex flow meters, which have a minimum Reynolds number below which the signal drops out entirely.

**RMS (Root Mean Square)**—The effective value of an AC voltage or current. A standard multimeter set to AC mode reads RMS values. Required for accurate measurement of AC input card threshold voltages.

**RTD (Resistance Temperature Detector)**—A temperature sensor that measures temperature by tracking the resistance change of a metal element (typically platinum, Pt100 or Pt1000). More accurate and stable than thermocouples, but more fragile and expensive. Available in 2-wire, 3-wire, and 4-wire configurations.

**Scaling**—The mathematical conversion from raw counts to engineering units (PSI, °F, GPM, etc.) using a linear formula. If raw counts match your measured current but the HMI reading is wrong, the problem is here.

**Shield Grounding**—The practice of connecting a cable's braided shield to ground at one end only (typically the PLC end) to drain electromagnetic interference without creating a ground loop.

**Sinking (Input)**—An input card configuration where current flows into the card terminal and out through the sensor to ground. Compatible with NPN sensors.

**SITVD**—The five-step troubleshooting method used throughout this book: Symptom, Isolation, Test, Verify, Document. A structured approach to prevent guessing and ensure you find the actual root cause.

**Snubber**—A device (sintered metal disc or orifice) that restricts flow to dampen pressure pulsations before they reach a pressure transmitter. Used in reciprocating pump and hydraulic applications. More dampening means slower response.

**Sourcing (Input)**—An input card configuration where the card provides voltage (+24V internally) and current flows through the sensor to the input terminal. Compatible with PNP sensors.

**SSR (Solid-State Relay)**—A relay with no moving parts that uses semiconductor switching. Faster switching than mechanical relays, no contact wear, but generates heat under load and can fail shorted (always ON).

**Strain Gauge**—A thin resistive element bonded to a surface that changes resistance when stretched or compressed. The sensing element inside load cells and many pressure transmitters.

**Structured Text (ST)**—A text-based PLC programming language similar to Pascal. Common in complex calculations, recipe management, and Siemens platforms.

**Terminal Block**—A modular connector used to join field wiring to panel wiring inside a control cabinet. The physical connection point you measure at when troubleshooting.

**Thermocouple**—A temperature sensor that generates a small voltage from the junction of two dissimilar metals. Self-powered (no excitation needed), rugged, and covers a wide temperature range. Common types: Type K (general purpose), Type J (lower temperature), Type T (cryogenic).

**Thermistor**—A semiconductor-based temperature sensor with a non-linear resistance curve. Common in HVAC and refrigeration, not in heavy industrial process control. Not interchangeable with RTDs or thermocouples.

**Triac**—A semiconductor switch used in some PLC output cards that can switch AC loads. Cannot reliably switch DC loads—a DC load on a triac output may latch ON permanently because the triac needs the AC zero-crossing to turn off.

**Ultrasonic Flow Meter**—A flow meter that measures velocity using sound waves. Clamp-on versions require no pipe penetration, making them the best portable verification tool for disputed flow readings.

**VDC (Volts DC)**—Direct current voltage. Industrial control circuits typically run on 24 VDC.

**VFD (Variable Frequency Drive)**—An electronic motor controller that adjusts motor speed by varying the frequency of the power supplied to the motor. Speed reference is often a 4-20 mA or 0-10 VDC analog signal from the PLC.

**Vortex Meter**—A flow meter that counts vortices shed by a bluff body in the flow stream. Frequency is proportional to velocity. No moving parts.

Dominant technology for steam flow measurement. Has a minimum flow threshold below which the signal drops out entirely.

**Wheatstone Bridge**—A circuit of four resistors arranged in a diamond pattern used to detect small resistance changes. The sensing element in load cells and strain-gauge pressure transmitters. A healthy bridge reads the rated resistance (typically 350 or 700 ohms) across both the excitation and signal terminals.

**Zero-Based Scaling** A scaling method where the minimum input signal maps to zero and the maximum input signal maps to the maximum raw count. For example, a 4-20mA analog input scaled 0-32767 in the PLC means 4mA = 0 counts and 20mA = 32767 counts. This differs from live-zero scaling where 4mA = a non-zero count (like 6400). Zero-based scaling is simpler math but loses the ability to detect sensor failures below 4mA.

**Zone Classification** The system used in North America (NEC) to categorize hazardous locations based on the likelihood of flammable gases, vapors, or dusts being present. Class I = gases/vapors, Class II = combustible dusts, Class III = fibers. Division 1 = hazard normally present, Division 2 = hazard only present under abnormal conditions. Electrical equipment must be rated for the zone where it's installed. For example, a sensor in a paint booth (Class I Division 1) requires explosion-proof housing. European systems use Zone 0/1/2 instead of Division 1/2, but the concept is the same: match the equipment rating to the hazard level.

# ABOUT THE AUTHOR

Robert Cummer spent thirty years troubleshooting industrial automation systems in automotive assembly, chemical processing, food manufacturing, and electronics production. He holds two engineering patents and has trained countless technicians and controls engineers in systematic troubleshooting methods.

Before founding Mangrove Publishing Company in 2025, Bob progressed from automotive test engineer to Technical Leader at Dow Chemical, where he specialized in instrumentation and control systems. His career spans VCR repair shops to Fortune 500 chemical plants, giving him a unique perspective on what works in the field versus what looks good on paper.

Bob developed the SITVD troubleshooting framework (Symptom, Isolate, Test, Verify, Document) after decades of pressure-filled situations where systematic thinking was the difference between hours of downtime and minutes of focused diagnosis.

The *Boots on the Ground* series reflects his mission: helping plant engineers, maintenance and instrument technicians think clearly under pressure through practical, field-tested guidance.

When he's not writing about instrumentation, Bob channels thirty years of real-world engineering into the Knox Ramsey thriller series—starting with *Dark Recipe*, a geopolitical techno-thriller built on the same first-principles thinking that drives his technical work.

He lives in Midland, MI, and is co-founder of Molly's Grape & Citrus LLC and Chief Technical Officer.

www.BootsOnTheGroundTech.com

www.ingramcontent.com/pod-product-compliance
Lightning Source LLC
LaVergne TN
LVHW020417070526
838199LV00055B/3647